W0114621

"The Brain Controls Everything"

A volume in
Cognition, Equity, and Society: International Perspectives
Bharath Sriraman, *Series Editor*

"The Brain Controls Everything"

Children's Ideas About the Body

Gunnhildur Óskarsdóttir

University of Iceland

INFORMATION AGE PUBLISHING, INC.
Charlotte, NC • www.infoagepub.com

Library of Congress Cataloging-in-Publication Data

A CIP record for this book is available from the Library of Congress
http://www.loc.gov

ISBN: 978-1-68123-378-9 (Paperback)
        978-1-68123-379-6 (Hardcover)
        978-1-68123-380-2 (ebook)

Copyright © 2016 Information Age Publishing Inc.

All rights reserved. No part of this publication may be reproduced, stored in a
retrieval system, or transmitted, in any form or by any means, electronic, mechanical,
photocopying, microfilming, recording or otherwise, without written permission
from the publisher.

Printed in the United States of America

Contents

List of Figures

The Brain Controls Everything, pages ix–xii
Copyright © 2016 by Information Age Publishing
All rights of reproduction in any form reserved.

List of Tables

The Brain Controls Everything, page xiii
Copyright © 2016 by Information Age Publishing
All rights of reproduction in any form reserved.

Foreword

This study explores how and under what circumstances children's ideas about the body change over the period of two school years, Primary 1 and 2 (6 and 7 years old), in a "normal" classroom setting in an Icelandic primary school. The focus is on children's ideas about the structure, location, and function of bones and other organs and how changes in pupils' ideas are affected by the curriculum, teaching methods, teaching materials, and teacher–pupil and peer interactions. Special attention is given to the differences between quiet children and more open children in respect to these issues. The theoretical background of the study is the constructivist view of learning and teaching with an emphasis on varied interactions as a pre-condition for learning and the importance of gauging children's initial knowledge on which to build their education. One class of 20 (19 in Primary 2) children took part in the research, along with their teacher and a sample of parents. A qualitative methodological approach was adopted although quantitative information was also obtained. Classroom observations and interviews were analysed by using elements from grounded theory and discourse analysis, and scales and pre-designed forms were used to analyse drawings, interviews, and diagnostic tasks. At the end of the project the children were generally more aware of the structures, locations, and functions of the various organs than they were of processes and how the organs were interrelated; and they were also more aware of the digestive system than other organ systems. The various teaching methods have different affects for different children and thus a variety of teaching methods are important

"The Brain Controls Everything", pages xv–xvi
Copyright © 2016 by Information Age Publishing
All rights of reproduction in any form reserved.

in order to maximize learning within a whole class. It is hard to conclude which teaching method is, overall, "the best one" although a combination of group demonstration, hands-on activities, information/telling, and discussion together were effective. Using drawings to get access to children's ideas can be very effective although young children may have difficulties in making drawings that represent their ideas. Furthermore, the imitation effect has also to be taken into account as drawings can present imitation rather than understanding; so other methods, such as interviews, should ideally be used as well. This research demonstrates the importance of the active engagement of the children as a group, but at the same time shows how the visible activity of individuals may not correlate with learning in the manner often presumed. The quiet children did not learn less than the others; indeed they learned more than the visibly active children. Although the study was undertaken (i.e., data collected) about ten years ago, its relevance and importance has not changed. The research makes a contribution to educational research in general as it is about teaching and learning, with an emphasis on the individual pupil within an educational setting and gives valuable insight into children's ideas about the body and how and under what conditions they develop.

Acknowledgment

The research was supported by Rannís (The Icelandic Centre for Research) Graduate Research Fund and the Research Fund of Iceland University of Education.

$$\underline{\quad\quad}$$
$$1$$
$$\underline{\quad\quad}$$

Introduction

For all my years as a primary school teacher and later as a lecturer and a teacher trainer at the Iceland University of Education I have been interested in the ideas young primary school children have about objects and in the dialogue they use to explain certain concepts and phenomena. I have also sometimes wondered about their understanding of these concepts and phenomena and in particular what kind of understanding lies behind the words or phrases they use. I have furthermore often thought about the multiplicity of ways in which the teacher has an influence and how her body language, her reactions to what the children say—for example, her eyes and her voice can have a great effect both to encourage but also to discourage the children in expressing their ideas (see Hayes 2004, p. 158–161).

Once I was teaching my 6-year-old pupils about air or rather exploring the concept of air with them. I asked questions like: What is air? Can we see air? How do we know that air exists? The children came up with a few ideas, such as: Air is everywhere. We cannot live without air. It comes into the mouth. You can sometimes smell it. We cannot catch it, it goes away. A balloon floats because it is full of air. Things fall down because

"The Brain Controls Everything", pages 1–6
Copyright © 2016 by Information Age Publishing
All rights of reproduction in any form reserved.

the air cannot hold them. We use air to blow out the light on a candle. These ideas are all more or less correct but the understanding that lies behind them can be vague. Ideas like these about air have made me even more interested in exploring children's ideas about certain concepts and phenomena in the environment and this interest of mine has motivated me to look into research that has been done on children's and young people's ideas about different things and phenomena especially in the area of science education (Driver, 1983; Driver, Guesne, & Tiberghien, 1985; Driver, Squires, Rushworth, & Wood-Robinson, 1994; Helldén, 1999, 2004a, 2004b; Holgersson, 2004; Reiss et al., 2002; Tunnicliffe, 2004; Tunnicliffe & Reiss, 1998; Tunnicliffe & Reiss, 1999b). These researchers in particular have stimulated my interest in the field and I build my work partly on their ideas. Driver (1983) talks about children's conceptual development and explores the ideas children have about a range of natural phenomena. Helldén's studies are longitudinal studies of how children's ideas about cycling of matter in the nature develop (Helldén, 2004b), and also of young people's ideas about their own development of biological knowledge (Helldén, 2004a). Holgersson's longitudinal study (2004) was on how children learn about nature and how their ideas develop, that is, how children react to and develop their ideas about candlelight and also other materials and processes involved in the change and conservation of materials. Reiss and Tunnicliffe have undertaken extensive research on children's ideas and understanding about the body.

But for those of us that are interested in education there are also studies that focus on teaching and the variety of methods teachers can use to find out about their pupils' ideas about different concepts and phenomenon (Black & Harlen, 1995; Carvalho, Silva, Lima, & Coquet, 2004; Helldén, 2004b; Naylor & Keogh, 2000; Norðdahl, 2004; Osborne, Wadsworth, & Black, 1992; Reiss & Tunnicliffe, 1999a; Tunnicliffe & Reiss, 1999a, 1999c; White & Gunstone, 1992). Osborne et al. (1992) and Black and Harlen (1995) write about the use of drawings, discussion, and different tasks used in the British SPACE research, *Science Processes and Concepts Exploration* (Black & Harlen, 1995; Osborne et al., 1992). Norðdahl (2004) used interviews with pairs of preschool children to explore their ideas about the transformation of materials in nature, and Helldén (2004b) also used interviews to get access to children's ideas about candlelight, circulation of water, and the decomposition of leaves. White and Gunstone (1992) on the other hand used concept maps to get access to children's ideas about different phenomena, and Carvalho et al. (2004) and Reiss and Tunnicliffe (1999a, 1999b; Tunnicliffe & Reiss, 1999a) also used drawings in their studies about the human body to find out about children's ideas. All these studies are of value but my particular interest is in those about the human body which is the focus of this study.

The *National Curriculum Guide: Natural Science* (Menntamálaráðuneytið, 1999) directs that pupils in primary two (seven years old) should know the main organs of the body, such as the heart, lungs, and stomach, and their functions. It is also stressed that children's conceptions of things should be taken into consideration when planning the curriculum, that is, account should be taken of their existing ideas about concepts and phenomena. For a number of years my BEd students have done an assignment (in pairs) where they are required to do an investigation into children's ideas about certain organs. They ask two children, one at a time, in primary one (six years old) about one or two organs in their body (or the skeleton), and they also ask them to draw a picture of the organ or organs they are discussing. In the light of the ideas presented by the children the BEd students plan four lessons where the focus is on the organs they asked about. Many interesting ideas emerged from these assignments that got me interested in exploring the field further. The children who were asked about the heart and its function almost all presented it as a V-shaped "Valentine's" heart; they also knew that if we did not have a heart we could not live. Many of those who were asked about the bones or the skeleton said that if we did not have any bones we would just be a "pile" and if we did not have any muscles we would be "lazy." The majority of the children drew the muscles on the upper arms and some of them also on the leg bones. Some said that the muscles made you "strong" and "busy" and you can "move houses" if you have muscles. You can also win the "strongest man in the world contest" if you have big muscles, an idea that is probably especially Icelandic, since Icelanders have won the contest several times and that is something Icelandic children know. The children's ideas about the lungs were also quite interesting. A majority knew that we breathe with our lungs. Some connected the nose to the lungs by saying that because we have two nostrils we have two lungs. One child said it was not good to have lungs. It was really disgusting! When asked if he did not have lungs he replied in horror: "No, of course not, you can only get lungs if you smoke!" Here the effect of the television and advertisements is obvious because there had been on the television a warning that showed how smoking can affect your lungs and make them black and unhealthy. I find these ideas young school children have about their body and its various organs—the structure of an organ or organ-system (what it looks like), its position (where it is in the body), its function (how it works), and process (e.g., the process of digestion)—both fascinating and also very important for a teacher to understand. There was never a doubt in my mind that this was an area I wanted to explore in a research project. I have also written, along with my colleague, Ragnheiður Hermannsdóttir, teaching material about the human body for the 1st–2nd year of the primary school, *Komdu og Skoðaðu Líkamann* (*Let's Look at the Human body*) (Óskarsdóttir & Hermannsdóttir, 2001a). The teaching material consists of (a) a "big book" with large pictures and a rich text for the teacher to use with the children, (b) a smaller textbook for the children, and (c) extensive material on the Internet—both information for teachers and ideas, activities, story-line approach, and interactive activities for children to do on their own or in pairs on the computer (Óskarsdóttir & Hermannsdóttir, 2001b). A well-known Icelandic artist, Sigrún Eldjárn, drew the illustrations in the book. This material is published by The National Centre for Educational Materials (NCEM) and is used in possibly all, Icelandic schools.

The paragraph in the *National Curriculum Guide: Natural Science* (Menntamálaráðuneytið, 1999), that recommends that the teacher take the children's conceptions and ideas into account when planning the curriculum and setting educational objectives is both important and valuable and should be emphasized. Many conceptual models have been made in order to formulate educational objectives. One of the best known is *Bloom's Taxonomy of Educational Objectives* (Bloom, 1956). According to Bloom and his associates learning is classified into three domains: cognitive, affective, and psychomotor where the cognitive domain includes objectives that are related to recognition of knowledge and the higher intellectual skills and abilities. According to them the cognitive domain is divided into six levels, that is, knowledge, comprehension, application, analysis, synthesis, and evaluation (Bloom, 1956). Cangelosi's (1992) modified model of Bloom's taxonomy is more simple as he divides cognitive objectives into two types of objectives (i.e., "knowledge level" and "intellectual level." Within the "knowledge level" it is distinguished between *simple knowledge* and *knowledge of process* and within "intellectual knowledge" between four types of intellectual learning: *comprehension, conceptualization, application, and creativity.* In this study a distinction is made between knowledge and understanding. Knowledge includes, for example, being able to state the structure and location of an organ or organ system and can be referred to as *simple knowledge* according to Cangelosi. Processes such as the steps involved in digestion would be referred to as *knowledge of process* according to Cangelosi, where pupils are required "to remember a sequence of steps in procedure" (p. 54). Understanding, however, means appreciating how things work, for example the function of an organ and how organs are interrelated. It can be difficult to distinguish between understanding of a function of an organ and how organs are interrelated within Cangelosi's four levels of "intellectual level" although this type of understanding probably best fits into *conceptualization* and *application*.

Extensive research into children's alternative conceptions has stimulated considerable interest in "constructivist" views of learning (Hodson & Hodson, 1998). Constructivism has been defined as:

> ...a theory of learning which holds that every learner constructs his or her ideas, as opposed to receiving them, complete and correct, from a teacher or authority source. This construction is an internal, personal, and often unconscious process. It consists largely of reinterpreting bits and pieces of knowledge—some obtained from first-hand personal experience, but some from communication with other people—to build a satisfactory and coherent picture of the world. (Selley, 1999, p. 3)

Thus, a constructivist approach examines how, when, and where the learners are active in the process of taking in information and building their own knowledge, of constructing (building) their own meaning, and learning (Cornu, Peters, & Collins, 2003). This calls for a discussion which involves teaching, the role of the teacher, teaching methods, teaching material, the classroom environment and classroom cultures, and interaction between the teacher and the pupils and between the pupils themselves. It also calls for definition of the term "active" and what "being active" means in this context. In the light of this, one can assume that there is a correlation between "being active" and learning and it is suggested here that this is worth examining.

My experience and my genuine interest in the constructivist view of learning and how it can be applied to classroom practice points the way ahead in undertaking my research and exploring this field.

1.1 The Structure of the Study

After the present chapter, the book consists of four chapters. In Chapter 2, "Review of Previous Research," the theoretical ideas about children's cognition, their ideas about the body; the importance of interaction in developing ideas, different studies relevant to the teaching about the body, and the notion of activity as related to teaching and learning are presented. The chapter concludes with the development of the research questions for the study. In Chapter 3, "Methodology," the design of the research is described: the participants and how several methods were used to obtain and analyze data about children's ideas about their body, and how their ideas change during the course of two school years. Chapter 4, "Results," presents the main findings of the study: the children's ideas about the body and how they changed, the effect of the teaching methods and teaching material used, and the effect of interaction in the classroom. In Chapter 5, "Discussion and Conclusions," the findings of the study are discussed with reference to the literature and the methods used, along with a discussion of practical implications and recommendations to enhance teaching and learning about the body.

2

Review of Previous Research

In this chapter selected literature is presented to provide a context for the rationale, the research questions, and the findings of the study. In Section 2.1, theoretical ideas about cognition will be addressed with special focus of constructivist perspective on teaching and learning. In Section 2.2, findings of other research on children's ideas about the body will be explored. In Section 2.3, the focus is on interaction in the classroom and how, and under what circumstances, children can learn from each other and the teacher. Section 2.4 addresses the complexity of teaching about the body, teaching methods, and the influence of teaching material. In Section 2.5, the focus is on the quiet, withdrawn children in the classroom and what the research literature says about their learning. The research questions of the study are finally presented in Section 2.6.

2.1 Theoretical Ideas About Cognition

Constructivism is based on the later work of Jean Piaget (mostly 10–15 years prior to his death) and Lev Vygotsky but also derives from the works of

"The Brain Controls Everything", pages 7–62
Copyright © 2016 by Information Age Publishing
All rights of reproduction in any form reserved.   **7**

other writers, notably John Dewey, Jerome Bruner, Howard Gardner, and Nelson Goodman (Fosnot, 1996). According to Piaget's constructivist theory, what we see, hear, and feel (our sensory world) is the result of our own perceptual activities and is therefore specific to our own ways of perceiving and conceiving. Knowledge, for Piaget, arises from perceptions, actions, and reflections and when Piaget speaks of interaction this does not only imply an organism that interacts with objects as they really are, "but rather a cognitive subject that is dealing with previously constructed perceptual and conceptual structures" (von Glaserfeld, 1996, p. 5). Thus, to Piaget interaction does not have to involve handling real objects but can also be something that the child is trying to solve or think about, building on ideas and experience that he or she already has.

Piaget worked as a biologist early in his life and he related his ideas about learning largely to biological ideas. He suggested that as children learn more about their environment they become better adapted to it. Piaget has called this process of adaptation "equilibration" which takes place when a person "assimilates" an experience and adjusts his knowledge through what he called "accommodation." According to Piaget the processes of assimilation and accommodation are necessary for cognitive growth and development. Equilibrium is the balance between assimilation and accommodation and is as necessary as the processes themselves (Piaget, 1977; Wadsworth, 1996). Although, for Piaget, maturation and various social and physical experiences are necessary for development, the essential and most important dynamic process is that of equilibration (Piaget, 1977; Driver, 1983). Whereas Piaget sought to study and illuminate the role of equilibration in learning, Vygotsky sought to study dialogue, that is, the dialogue between the individual and his social environment, the interaction that takes place, the language and the culture that the individual is a part of. He was interested in the learning of concepts, but also in the role of the adult (e.g., the teacher) and the learners´ peers as they discuss, question, explain, and negotiate meaning (Fosnot, 1996; Vygotsky, 1986). The Piagetian view holds that action is primary, "with language and other ways of using symbols following from the development of more general, underlying cognitive structures" (Hodson & Hodson, 1998, p. 36). Vygotsky (1987), on the other hand, believed that language was a tool for problem solving, the greater the problem the greater the importance of speech. Vygotsky was critical of the Piagetian view that learning must necessarily lag behind development, that is, that learning occurs only and when appropriate cognitive structures are in place to enable experience to be appropriately represented and acted upon. Although Vygotsky agreed with Piaget in that children function according to norms of development, according to Vygotsky it is incorrect to say that they function only at a particular level. He believed that with appropriate assistance children aged 8 can solve tasks and problems at the level of 9- or even 12-year-old children:

> We assist each child through demonstration, leading questions, and by introducing the initial elements of the task's solution. With this help or collaboration from an adult, one of these children solves problem characteristics of a twelve year old, while the other solves problem only at a level typical of a nine year old. (Vygotsky, 1987, p. 209)

Crucial in this context is Vygotsky's concept of the "Zone of Proximal Development" (ZPD). Vygotsky explains:

> The Zone of Proximal Development is the distance between the actual developmental level as determined by independent problem solving and the level of potential development as determined through problem solving under adult guidance or in collaboration with more capable peers. (Vygotsky, 1978, p. 113)

Piaget's version of constructivism sought to identify the structures of mind and cognitive characteristics of each stage of development (Noddings, 1995). Many constructivist educators have criticized Piaget's work for concentrating too heavily on the individual child's interaction with and manipulation of objects but neglecting the social dimension of learning, referring to the point that most of us learn by interaction with others (Noddings, 1995). Because of these differences, terms such as "cognitive constructivism" and "social constructivism" have been common in the literature, each term depending on the emphasis on either the cognitive (Piaget's) or social (Vygotsky's) view of constructivism (Cole & Wertsch, 1996; Fosnot, 1996; Noddings, 1995). Although the difference between the two main strands of constructivism may be quite substantial—in that Piaget's approach looks at cognitive development as a sequence of relatively distinct stages, each with its own set of limitations and abilities, while Vygotsky's approach focuses more on the social conditions that facilitate cognitive development—the differences between their ideas may have been over-emphasized. Daniels (2001) discusses the issues which serve to distinguish between Piaget's and Vygotsky's theories and the overemphasis that has been on the difference between them and says that there should be no need for an "either–or" position on the Piaget–Vygotsky "debate" (Daniels, 2001, p. 38). Issues and elements from both theories can easily go together and support each other. Other writers have also pointed out that there are many similarities in their work (Cole & Wertsch, 1996; Daniels, 2001; Kozulin, 2003). There are several features that characterize both theories, namely: active thinking and learning, the importance of relating new information to prior knowledge, the concept of readiness, the importance of interaction, differences between the thought processes of children and adults, and the role of language (Ormrod, 1995). In principle, Piaget did not deny the role of the social world in the construction of knowledge and Vygotsky did insist on the centrality of the active construction of knowledge in an active child and an active environment (Cole & Wertsch, 1996). Cole and Wertsch (1996) emphasize, as Vygotsky did, that the element of culture is essential in the process of the construction of knowledge. It can be the culture of the society the child lives in, the culture of the home or the culture of the school:

> This follows from the fact that the artifacts which enter into human psychological functions are themselves culturally, historically, and institutionally situated. In a sense, then, there is no tool that is adequate to all tasks, and there is no universally appropriate form of cultural mediation. (p. 253)

They also point out that for Vygotsky, like Piaget, the relationship between the individual and the social is necessarily rational although "social origins take on a special importance in Vygotsky's theories that is less symmetrical than Piaget's notion of social equilibration as resulting from the interplay of the operations that enter into cooperation" (p. 254).

There are other similarities that can be identified and in many cases the strengths of the theoretical frame which one proposes can compensate for the weaknesses of the other. One aspect which becomes important in the present study is the power of egocentric speech: both Piaget and Vygotsky were interested in the power and function of egocentric speech in childhood even though their ideas about it are not the same. Piaget was the first to recognize the special function of egocentric speech in the child and to understand its theoretical significance and according to him the child's egocentric speech is a direct expression of the child's thought (Ormrod, 1995; Vygotsky, 1986, 1987). Piaget sees thought driving on language where concepts are developed and learned which allow the language to grow. To Piaget egocentric speech is something that comes before and then develops and leads to and is replaced by social speech—that is, egocentric speech goes away with maturity. Vygotsky, however, put social speech first whereby egocentric speech is internalized to become verbal thought where language guides and drives thought on. According to Vygotsky, thought and language are separate functions of very young children, infants, and toddlers, and in these early years thinking occurs independently of language; at first, language is used by the young child primarily as means of communication rather than a mechanism of thought, but around the age of two, thought and language become intertwined as children begin to express their thoughts when they speak. At this age children also begin self-talk which Piaget interpreted as egocentric speech. Gradually self-talk becomes what has been called "inner speech," where children talk to themselves covertly rather than overtly and guide and direct themselves in much the same way that adults have previously guided them, which means that they begin to provide their own "scaffolding" (Ormrod, 1995). For young children (preschool children), "egocentric speech is the attempt to make sense of the situation in words, to find a solution to a problem or plan the next action" (Vygotsky, 1987, p. 70). Older children (school-age children) act somewhat differently as they view the situation, think, and then try to find the solution (Vygotsky, 1986, 1987).

Although constructivism is a theory about learning and knowledge, it is not a theory or prescription for teaching. It suggests, however, an approach to teaching that gives the learner the opportunity to become engaged in an activity, discourse, and reflection (Fosnot, 1996). Vygotsky emphasized that teachers should put their effort into the ZPD because it advanced learning and development and it is in relation to the ZPD that the term "scaffolding" is first used (Wood, Bruner, & Ross, 1976). Bruner developed the idea of "scaffolding" in the light of Vygotsky's theory of social constructivism. Teachers, parents, and peers can scaffold a child's understanding through a problem-solving process (Bruner & Haste, 1987). Some such problem solving involves scaffolding situations, in which a more knowledgeable person helps a less knowledgeable one to learn. Scaffolding occurs in the context of parents helping their children, teachers helping their students, coaches helping their players, and more advanced learners helping the less advanced ones (Siegler, 1996; Wood et al., 1976). Scaffolding is a technique that supports children's learning in the same way as when builders creating a new building have to construct an external structure, or scaffold, around the building. This scaffold provides support for the workers, a place to stand on, until the building itself is strong enough to support and hold them. When the building becomes self-supporting the scaffold becomes less necessary and is gradually taken away (Ormrod, 1995).

Like the builder using a scaffold, the teacher guiding a child through a new task may also provide an initial scaffold to support the child to do the task. Teachers provide scaffolding when, for example, they give a child hints about how to tackle a new problem-solving task. As children become more confident and more used to performing such tasks, the guidance is gradually phased out and the children will eventually become more skilled and can perform those tasks on their own. Depending on the task and the particular child, the teacher can provide a variety of support to help the child master tasks within their ZPD. As the children become more confident the teacher can gradually withdraw the scaffold and allow the children to perform the task independently: individually, in pairs, or in a group (Ormrod, 1995). Thus, scaffolding and activity are interrelated and it depends on the child and its skills and understanding how much and what kind of scaffold is needed for each activity.

Many studies have probed the ways in which teachers who are committed to a constructivist philosophy construct teaching and learning (Cornu, Peters, & Collins, 2003; Fosnot, 1996; Hodson & Hodson, 1998; Noddings, 1995; Ogden, 2000; Porter & Harwood, 2003). Some teaching approaches emphasize this approach more than others. Constructivist teachers do not generally favor lecturing and telling but emphasize instead the active engagement of pupils in establishing and pursuing their own learning objectives and active learning, in which the pupils are engaged in action, discussion, and reflection which gives the teacher a better opportunity to know what the pupils are thinking in order to facilitate their learning (Noddings, 1995).

Constructivist teachers often use methods adapted from Piaget where they begin a topic or a lesson by asking pupils to express what they think and then follow up with prompts, challenges, and questions about the general usefulness of the methods the pupils have chosen (Noddings, 1995). Here the teacher is seen as a provider and manager of a suitable environment in which learning subsequently occurs. Vygotskian theory, however, gives the teacher the central role of leading the children to new levels of conceptual understanding by interacting and talking to them, and learning is seen through guided and modeled participation (Hodson & Hodson, 1998). Both views aim at getting children active but in a different way: The children's involvement is the key to their being engaged (Richardson, 2006). But what does "being active" mean in the context of learning in the classroom? According to Maynard (2001), teachers, when referring to the importance of active learning, seem to be implying that children need to "engage" with curriculum content and the activity therefore has to be linked to the children's present understanding and be seen as purposeful by the children (Maynard, 2001). Her study showed how the teachers maintained that in their learning children needed to be given a certain amount of information or knowledge and then encouraged to react. Learning was seen as an active and interactive process, and the teachers said they constantly had to alter their approaches according to the children's reaction in order to ensure understanding. One teacher in the study explained active learning so:

In order for anybody to learn, something comes in and connects to something and by coming out again in some form, writing, modelling, drawing, talking, you are reaffirming (learning) to your brain. (p. 46)

The student teachers that were also involved in this study interpreted "active learning" in different ways. The term "active learning" meant for many students simply "doing practical work." Active learning therefore is often understood as being physically active doing something. Other students saw active learning just as "looking things up in books." There were also students who thought that the teacher should not intervene at all, and this was linked to the view that active learning was "natural," especially for younger children (Maynard, 2001). According to Kyriacou (1997) active learning typically involves activities such as small group work, project work, role play, problem-solving activities and investigations, and computer-assisted learning tasks, where a high degree of control over the learning process is given to the pupils (Kyriacou, 1997). Being active is what causes children both physically and cognitively to develop and construct their own view of the world around them and to personalize the experience and to apply it in a way that makes sense to them individually (Bruner & Haste, 1987). According to this view, "being active" does not necessarily mean physically (i.e., visibly) active. A child can be an active listener and thus be active watching a demonstration, pictures, or a video, or listening to a conversation or an explanation. Children can be active in a different way depending on the activity or topic, whether taking part in discussion, listening to the teacher or the views of others, or taking part in or just watching a demonstration, or as Wilson and Myers (2000) put it when defining learning as an active participation:

Learning is seen in terms of belonging and participating in communities of practice. Learning is seen as a dialectical process of interaction with other people, tools, and the physical world. Cognition is tied to action—either direct physical action or deliberate reflection and internal action. To understand what is learned is to see how it is learned within the activity context. (p. 71)

Thus Wilson and Myers are suggesting that children can be active in very different ways. Active learning can include a variety of actions that can be physical or internal: The child can be working alone or in collaboration with others and be participating in both cases, both "physically" active in doing something and also active "cognitively" by thinking and reflecting.

The "social constructivist" perspective suggests that the most effective form of learning is inquiry-oriented, personalized, and collaborative: a creative, problem-solving activity that the child works on, whether alone or in collaboration with others. The inquiry can be either literature- or media-based or it can be a hands-on or field-based inquiry (Hodson & Hodson, 1998). Hodson and Hodson have identified inquiry as having five stages from the social constructivist teaching perspective: *ignition, design and planning, performance, interpretation and reporting*, and finally, *communicating*. They emphasize that the teacher participates extensively and supportively in all the inquiry activities. They also emphasize that inquiry-based learning is not designed to lead students to a predetermined view or solution. They stress that teachers must include all student contributions to class discussion at all stages. Through collaborative experiences and talk, social interaction becomes an internal process, and teachers can focus on the ZPD and enable each child to learn from the teacher's more expert knowledge:

> In this way, conceptual knowledge and procedural knowledge that is first encountered, discussed and mastered in social interaction with peers and teacher becomes part of the student's personal framework of understanding. (Hodson & Hodson, 1998, p. 40)

There has been concern about how Vygotsky's ZPD is interpreted. One of those who has expressed concern is Carugati (1999) where she says:

> The Vygotskian approach is concerned with an overwhelming emphasis on the transmission from adult-expert to child-novice in the way of thinking, with no reference to the possible role played by peers. (p. 217)

According to Carugati, in a discussion children might benefit from their partner's answer even when it is inferior to their own. This means that it is not necessary for one partner to be more expert than the other; and this is of a major importance for a reformulation of the understanding of what ZPD might be:

> What first seems to be strange, mysterious, troublesome, meaningless for children in isolation, could become familiar, and meaningful through children's participation in their peer culture (both in nursery and in elementary school), where discussion, play, provisional interpretations of the bizarre adults' world are mediational tools built up for interpreting and reproducing the adults' world. (p. 225)

Kutnick (2001) also discusses what he calls a "naive application" of Vygotsky's work to the educational field. He criticizes some of the current interpretations concerning the ZPD with the great emphasis on adult–child interaction as it has been interpreted and used in teaching and learning situations. He uses results from different studies to support his view that a sharing among equal partners can form a basis for learning (Kutnick, 2001).

According to Kutnick (2001), Vygotsky's theory about ZPD incorporates a situational definition to reinforce the existence of the asymmetrical relationship. He claims that "this definition explains an asymmetrical teacher–pupil dynamic that characterizes the control of knowledge in society (and classrooms) but excludes alternative, co-operative relationships that may also promote learning and development" (p. 82). According to Kutnick cooperative or symmetrical relationships are often not taken into account as a site for ZPD because peers do not often show the emotional security that will allow them to "learn together" which is thought to be an important condition for learning to take place. Vygotsky himself, however, was aware of the relational basis of ZPD but was also aware of that this kind of relationship could take place among peers as well as expert and novice (adult or child), although this interpretation has not been made evident in the dominant writings of Vygotsky (Kutnick, 2001). Kutnick stresses the fact that Vygotsky also used an "expanded" ZPD involving both hierarchical and mutual interpersonal relationships and emphasized the need to draw upon both kinds of relationships to support learning within the ZPD (Kutnick, 2001).

Daniels (2001) discusses how Vygotsky's theory is presented in different ways and how different approaches to translations of his work may give rise to fundamentally different interpretations. Daniels, like Cole & Wertsch (1996) and Kozulin (2003), talks about the importance of cultural, historical and social influences as a basis of individual development in Vygotsky's ideas where cognition is seen to be situated in specific social, cultural, and historical circumstances (Daniels, 2001, p. 39). He suggests that "the final phase of Vygotsky's work suggests the need to move towards a broad analysis of behavior and consciousness which articulates and clarifies the social cultural and historical basis of development" (p. 32). Cultural development, he continues, involves the relationship between both the "social" and the "individual"; he uses the term "mediation," which suggests that there is not necessarily a simple directional relationship from the social to the individual, rather there is a dialectical relationship between the two (Daniels, 2001).

But how should the teaching environment be developed and what resources should be available to encourage both social and individual development? And how does this apply to the "constructivist classroom"? According to Gould (1996) there are several characteristics that so-called constructivist classrooms and schools share:

> They focus on big ideas rather than facts; they encourage and empower students to follow their own interests, to make connections, to reformulate ideas, and to reach unique conclusions. (p. 93)

Gould says, however, that we have to take into account the great number of children in each class and their different backgrounds and experiences, and be aware of the social and cognitive dimension of the classroom activities that the teacher provides. Harlen (2000) stresses that the teacher has a vitally important role in the success of learning in general, not least in the constructivist classroom, and stresses that the teacher must establish a supportive and positive classroom environment where the children feel free to ask and answer open questions and participate in discussion (Harlen, 2000).

Although formalized subject matter concepts can be learned at school they do not become meaningful for the child until they become active in the child's life (Hedegaard, 1999 p. 30). Therefore school teaching must connect the subject matter concepts with everyday concepts in a way that widens and develops children's abilities. This is also supported by Reiss' (2000) results from his five-year longitudinal study into pupils' learning in science where he suggests that school science education can only succeed when pupils believe that the science they are being taught is of personal worth to themselves, noting that "personal worth" must not be construed narrowly.

According to Hedegaard (1999) there can be qualitative differences between thinking procedures according to which type of knowledge is available and even within the same subject area the same person can use different forms of knowledge and procedures dependent on his motivation and social conditions. Theoretical knowledge can combine with everyday concepts and thereby help the child overcome the gap between knowledge and thinking within and outside school.

According to this, school teaching has to keep in mind how knowledge conceived at school can be transformed into active knowledge since, as Vygotsky (1986) has pointed out, the formal abstract knowledge that children learn at school does not become active until it becomes functional in the child's life.

Thus, central to the constructivist view of learning is the notion that the learner is active in the process of taking in information and building knowledge and understanding and that the learner can only make sense of new knowledge in terms of his or her existing understanding (Cornu et al., 2003; Keogh & Naylor, 1996).

The learner and the teacher, very explicitly, share a responsibility. The teacher is like a guide that has to lead the learner along the way ahead. But before she does so she has to know where the learner is situated, otherwise moving ahead would not make much sense and the learner could get lost. The aim is therefore to move the learner further by planning learning activities which challenge the learner's existing ideas (Selley, 1999).

According to Keogh and Naylor (1996) constructivist perspectives have had a significant impact on recent research in science but most of this has been concerned with finding out children's ideas in order to inform teaching. They believe that the model of constructivism that is generally put forward and that has been discussed here is based largely on methods that work for researchers but not so easily for teachers in typical classrooms. It can be difficult to find out the ideas that all the children in a class of 6 or 7 year-olds have about certain issues and plan an appropriate curriculum for every individual, and there has been some criticism of how the constructivist view is presented and practiced in science education.

Osborne (1996), in particular, offers a critique of constructivism in science education saying that constructivism does not recognize its own limitations especially regarding pedagogy where there has been overemphasis of the construction of the concepts, either personally or through discourse: "constructivist pedagogy often makes a fallacious connection between the manner in which new scientific knowledge is created and the manner in which existing scientific knowledge is learned" (p. 54). According to Osborne, constructivism has focused very strongly on the learner's beliefs and the social construction of reality, but has failed to look at the manner in which knowledge is constructed by the learner. He says that "constructivists have forgotten that it is the world that imposes constraints on human thought, and not the human thought that imposes constraints on the world" (p. 77). He also discusses how constructivism has been applied to the process of teaching and learning and refers to Solomon (1994), who identified what Osborne calls "the greatest success of constructivism" by the shift in describing pupil errors from "mistakes" to "misconceptions" or more often "alternative frameworks" and therefore changed the common and the unremarkable to something significant (Osborne, 1996 p. 63).

Osborne also argues that constructivists in science education have chosen to emphasize cooperation and discussion-based activities in order to promote knowledge but ignore the fact that there could be a place and role for telling, showing, and demonstrating. This was also pointed out by Noddings (1995). Evidence from research indicates that the approach which constructivist pedagogy offers to learning is preferred by some students but not all and, likewise, is effective for some students but not all. "Individuals vary in their preferred learning style and no one model will meet the diverse needs of different children" (Osborne, 1996, p. 75–76) and there is no one teaching strategy that will achieve success with all pupils. Consequently, science education should consist of a wide variety of teaching methods and take account of the fact that each individual learner is unique and also uniquely determined. Therefore an improved science education has to adopt a pedagogy which values variety and diversity (Osborne, 1996).

What Osborne is discussing here is an important contribution to the literature on constructivist teaching and learning. In the constructivist classroom there has to be a place for cooperation and discussion-based activities and also a place for a wide variety of teaching methods like telling, showing, and demonstrating and it has to be taken into account that each individual learner is unique. "Being actively engaged" in the classroom can mean an active listener or an active watcher. It can mean active interaction within oneself, with peers, with the environment, or with the things being explored. So when and where learners are active by virtue of being in the process of taking in information and building knowledge and understanding, namely, constructing their own learning, that is the activity which constructivism emphasizes. Culture has a great impact on what is processed and taught and then learned in the classroom, a view discussed by Cole and Wertsch (1996) and Kozulin (2003), but they emphasize Vygotsky's idea that culture influences cognitive development. According to Cole and Wertsch (1996) it is impossible not to be socioculturally situated when carrying out an action, or as they put it: "Conversely there is no tool that is adequate to all tasks, and there is no universally appropriate form of cultural mediation" (p. 253). And this should not be overlooked as each culture has its own psychological tools and situations in which certain tools are appropriated. These tools can be symbolic artifacts like signs, symbols, or texts that help individuals to master their own perception, attention, and memory (Kozulin, 2003, p. 15–16). According to Kozulin:

> The process of appropriation of psychological tools differs from the process of content learning. This difference reflects the fact that whereas content material often reproduces empirical realities with which students become acquainted in everyday life, psychological tools can be acquired only in the course of special learning activities. (p. 25)

For example, if the children are learning about Italy and the content knowledge that Rome is the capital of Italy, this corresponds to empirical reality and can be learned by students, both spontaneously in their everyday life or as a part of school curriculum. On the other hand, a symbolic tool like a map of Italy, including Rome, can help the students to apply, remember, and understand better the content knowledge (Kozulin, 2003). This would also apply to teaching and learning about the human body where the content knowledge is that, for example, the heart is an organ that we cannot live without and a muscle that pumps blood all around the body. Here a picture or a model of the heart could also help pupils to apply, remember, and understand better the structure and function of the heart.

To sum up, constructivism is a theory about learning which includes getting children to be active in their own learning because their involvement is essential for learning to take place. In the literature the terms "active" and "involvement" in the context of learning in the classroom are both described as being "physically" active in doing something and also active "cognitively" by thinking and reflecting. Children can be active in different ways depending on activity, topic, or situation. Vygotsky's and Piaget's contributions to constructivism, often called "social" and "personal," have important issues and strands in common that should be used to support constructivist pedagogy with all the strengths involved in both strands rather than putting effort into emphasizing all the differences and putting them against each other as "either–or." It is as if Vygotsky's ideas about the importance of talk had been taken out of context and used by some curriculum developers to support their own view and to bring about changes in their own favor. Too much effort has been put into looking at individual interaction with objects to the neglect of the social dimension of learning in the ideas of Piaget.

2.2 Children's Ideas About the Body

A number of studies have been undertaken on the understanding and the development of children's ideas about scientific concepts (Carey, 1985; Driver, Guesne, & Tiberghien, 1985; Driver, Squires, Rushworth, & Wood-Robinson, 1994; Lawson, 1988; Osborne & Freyberg, 1985). Children come to school with ideas and interpretations concerning certain concepts or phenomena even though they have never had any formal instruction on these concepts whatsoever. Children form their ideas and interpretations on the basis of everyday life and experience (Driver et al., 1985). According to Farmery (2002), children build up "scientific" knowledge from a range of sources outside the school environment:

> ...knowledge that may be very different from that which we would wish them to develop. These different understandings are often referred to as pupils' misconceptions. (p. 103)

Some misconceptions are quite common and may be very resistant to change. It is therefore important for the teacher to be aware of them and be able to respond appropriately when they occur (Farmery, 2002).

Gellert (1962) made a study of the ideas hospitalized children aged 4–16 had about the human body. The dominant answers in Gellert's study to her question of what is inside people were similar to the results of the SPACE study (Osborne, Wadsworth, & Black, 1992) The youngest children in Gellert's study (5–6 years) said there were food, blood, and bones and some thought of the body as a container with blood and food. The 7 and 8 year olds also mentioned the heart and the 9 and 10 year olds added several organs to the list. The children were interviewed and asked to draw specific major organs and asked about their functions. Her study showed that children held a number of common misconceptions about the body, such as the lungs were located in the neck or in the head and that the heart helps us to breathe. She also found out that children's knowledge about the body seemed to increase sharply at around the age of nine (Gellert, 1962).

Carey (1985) reviewed a number of studies on children's ideas about the body. According to her it is not until the age of 10 that children appear to understand that the body contains a number of organs which function together so we can live. It is not until then that "they know that the circulatory, respiratory, and digestive systems are related, and they are beginning to conceptualize the relations between the brain and the body" (p. 51). Children do not seem to grasp the function of internal body parts until then, but 4-, 5-, and 6-year-old children are familiar with the external body parts, like legs, feet, arms, nose, eyes, ears, and hair. These young children thought that consumption of anything, including water, would lead to the body gaining weight and they also thought that some diets were healthier than others and better for your growth. By the time the children were 8 years old most of them started differentiating between the different kinds of food that could make people fat or strong. Carey calls the ideas the younger children hold "psychological" ideas in contrast to "biological": according to her by the age of 9 or 10 children understand more in terms of biological principles and she describes how intuitive biology emerges from an intuitive psychology between the ages of 4 and 10. She describes "intuitive psychological" ideas as being based on "intentional causality" and says they are psychological because they are explained in terms of beliefs. When children try to understand human behavior they catalog and classify and make generalizations about its consequences. Specific consequences include, for example, gaining weight from eating too much candy. General consequences involve approval and punishment and its causes in terms of wants and beliefs. According to Carey, eating, growing, sleeping, breathing, dying, the heart beat, and having babies are all assimilated into this framework. It is, for example, enough to explain the function of the heart by saying that "it ticks" (referring to its behavior), although some children see the heart as the organ of the emotions and of morality. The causes and consequences of death and growth are seen in terms of human behavior (Carey, 1985).

In an attempt to distinguish the thoughts that quite young children (below the age of 9–10) have about human bodies, Carey (1985) writes about the "psychological ideas" that these children have. Carey provides no clear definition of the term but uses it to mean that young children attribute intentions not only to whole persons but to their body parts, so that the heart has to beat and the stomach has to digest food:

> ... the stomach does not want to digest the food; the heart does not believe that circulating the blood will distribute food throughout the body. That's just the way the body works. (p. 69)

She says that by the age of 10, children have another framework for understanding all of these behaviors which is now in terms of the integrated functioning of internal body parts, where what she calls "intentional causation" or "psychological ideas" play no role when explaining the working of the body (Carey, 1985).

There has been some criticism of this view held by Carey and some studies show that children possess biological knowledge much younger than Carey suggested and alternative proposals about the development of children's intuitive biological theory have been advanced, for example by Hatano and Inagaki (1994; Inagaki & Hatano, 1993). The results of their research suggest that by the age of 6 children have acquired a form of biology as an autonomous domain which is separate from that of psychology and that this domain is structured around different causal principles from those that structure the intuitive biological theory of older children and adults (Hatano & Inagaki, 1994; Inagaki & Hatano, 1993). One of their studies indicated that children as young as 6 years of age understand the distinction between mind and body and they suggest that 6-year-old children

> possess a form of biology which is applied to internal bodily functions; they not only understand the distinction between mind and body but also have a causal explanatory framework for bodily processes, which is not person-intentional or psychological, and might be called vitalistic. (Inagaki & Hatano, 1993, p. 1536)

They hypothesize that even young children are able to give vitalistic causal explanations for biological phenomena and that 6-year-old children prefer vitalistic explanations to intentional explanations about bodily functions. According to them, children who are reluctant to rely on intentional causality for biological phenomena, but are not yet able to use mechanistic causality, often rely on an intermediate form of causality which they call "vitalistic" when explaining bodily processes. A standard understanding of vitalism is according to the Skeptic's Dictionary (Caroll, 2005): "Vitalism is the metaphysical doctrine that living organisms possess a nonphysical inner force or energy that gives them the property of life." This is how I understand it, namely that vitalist children (and adults) talk as if organs can act independently and even have their own "mind."

Inagaki and Hatano (1993) argued that the development in thinking about biology is the shift from a vitalistic to a mechanistic explanatory stage. They provided examples: A vitalistic response to the question "Why does blood flow to different parts of our bodies?" could be "Because our heart works hard to send out life and energy with blood," compared to a mechanical response "Because the heart sends the blood by working as a pump." The intentional response to this question could be "Because we move our body, hoping the blood will flow in it" (p. 1540).

According to Inagaki and Hatano, older children reject intentional explanations for biological phenomena and use mechanistic explanations (Hatano & Inagaki, 1994, 1997; Inagaki & Hatano, 1993). Jaakkola and Slaughter (2002) agree with Hatano and Inagaki and in their own study look for evidence of "biological teleology" (teleology with a biological goal) or a "causal principle," because according to them, when speaking of children, "knowledge of organ function might be intertwined with their understanding of biological teleology, as they do construct the notion of a bodily machine" (Jaakkola & Slaughter, 2002, p. 328). They found that children's general factual knowledge about the body and the knowledge about organ location and organ function increased steadily between the ages of 4 and 8, and they demonstrated that children between 4 and 6 years of age begin spontaneously to refer to "life" as the purpose or goal of human bodily function. Results from their studies also show that by the time the pupils they studied were eight years old they mostly had a broad knowledge of the internal structure of the body and were aware of a wide variety of organs although they did not know how the organs were connected or how they were part of an organ system (Jaakkola & Slaughter, 2002). A study by Slaughter and Lyons (2003) also looks into the development of vitalistic reasoning in young children's (4–6 years old) concepts of life, the human body, and death. They suggest that the acquisition of a vitalistic causal explanatory framework for understanding body function organizes children's knowledge and directs further learning. The results show that the Australian children in their study, like the Japanese children in Inagaki and Hatano's study, adopted a vitalistic mode of construing when reasoning about the human body (Slaughter & Lyons, 2003).

Miller and Bartsch (1997) undertook a study of children aged 6 and 8 and adult college students, where many questions were adapted from Inagaki and Hatano (1993) in order to provide a meaningful comparison to their work. Their results suggest that young children (6 years old) think of biological phenomena as being qualitatively different from psychological phenomena and thus warranting different sorts of causal explanations, as Inagaki and Hatano (1993) claim. According to Miller and Bartsch (1997) the children in their study "did favour vitalistic explanations more for biological phenomena in general than for either psychological or mechanical phenomena, which would be consistent with supposing that they regard vitalism as a causal framework special to biology" (p. 163–164). Miller and Bartsch also claim, on the basis of their findings, that children do conceive biology as a domain different from psychology. They also claim, and this is different from Inagaki and Hatano's studies, that children's attribution of intention to biological organs or body parts (i.e., biological attribution) did not differ from the adult college students in their study, so according to them children's thinking about biology is not more vitalistic than adults.

An extensive study, the English Primary SPACE Project (*Science Processes and Concepts Exploration*), on children's conceptions on scientific concepts, showed that the children in the study were operating with a knowledge based on simple broad mechanisms like, "you need food to keep you alive" and "blood keeps you alive" which suggests that children's vitalistic explanations were seen by them as being comprehensive and adequate since many were correct (Osborne et al., 1992). The study also showed important differences between "correct" ideas and clear misconceptions, using a variety of different methods, for example discussion, log books (free writing and drawing), drawings, and individual discussion (Osborne et al., 1992). The children in the SPACE study were grouped by age into primary (5–7 years old), lower juniors (8–9 years old), and upper juniors (10–11 years old). The youngest children (5–6 years old) were able to name the parts of the body that they could see, touch, or hear and they could draw them. Furthermore, the study showed that children draw the organs that are more easily sensed, like the heart, which beats, and the bones, which they can feel. Some of the children in the SPACE study thought of the body like an empty container filled with blood and when they cut themselves the blood would simply flow out. Some of them also tended to think of the body as a container which contains different unconnected organs, bones, and blood (Black & Harlen, 1995; Osborne et al., 1992). In the SPACE research, the predominant organs named by all children were the heart, bones, stomach, and brain. The study also revealed that many children were not aware of the correct size or the location of the organs which is probably because the internal organs are not visible or touchable and therefore it is difficult for the child to develop knowledge of their size and correct location. Organs that are not part of everyday language like kidneys, liver, intestines, and even lungs were usually excluded by the children although most of the children knew that we need air to live but very few were able to locate the lungs on a drawing of the human body (Osborne et al., 1992).

A study undertaken by Cuthbert (2000) showed that 8–9-year-old children seemed to produce similar results as those of the 8–9 year-olds in Osborne et al. (1992). Children's drawings were analyzed and the majority of children drew isolated and unconnected organs in their body maps. Most children drew hearts and veins but very few connected the veins to the heart in their drawings and the majority of them drew brains but very few included nerves. A large number of children included various bones and placed them more or less in their correct position but the children did not include ligaments or joints (Cuthbert, 2000).

A study involving 586 pupils aged either 7 or 15 years old from 11 different countries, where children were asked to draw what they thought was inside themselves showed that the digestive system, the gaseous system, and the skeletal system were generally the best drawn organ systems. But very few drawings showed the muscular system, the endocrine system, or the circulation system. The analysis also showed that 7-year-old children frequently had a broad knowledge of their internal structure and also were aware of a wide variety of organs although they were not aware of how organs were interrelated or connected as parts within an organ system (Reiss et al., 2002).

In the light of the studies that have been discussed here, one can agree that biological knowledge develops between 5 and 10/11 years of age (Carey, 1985; Osborne et al., 1992) which, however, is a wide age range. It seems though that from the age of 4 there are gradual changes in children's ideas (Hatano & Inagaki, 1997; Teixeira, 2000; Toyama, 2000) and by the age of eight children have developed a broad knowledge of the internal structure of the body and are aware of a variety of organs although they do not know or understand how they are connected (Jaakkola & Slaughter, 2002; Reiss & Tunnicliffe, 2001). In the SPACE study, the major difference in biological knowledge existed prior to the intervention, which supports the case that there is some development in children's biological knowledge during the transition from Age 5/6 to Age 8/9 and this is reflected in the data (Osborne et al., 1992).

According to Gellert (1962), children's knowledge about the body seems to increase sharply at around the age of nine. Hatano and Inagaki (1994), however, suggest that at the age of 6 children know of internal bodily functions and have acquired a form of biology as an autonomous domain, and also have a causal explanatory framework for bodily processes which might be called "vitalistic," as discussed above.

Whatever the exact age, children's ideas about the body seem to develop extensively from the age of 4/5 to 8/9. Younger children seem to be aware of their external body parts and the main organs which are, according to the different studies, bones, blood, heart, and brain. The children see the bones and other organs at first as isolated parts but then gradually start to put bits and pieces together and see them as systems, but it is not until the age of 10/11 that they seem to understand some of the processes and systems of the body, while studies indicate that excretion is possibly a relatively poorly understood process by children under 11 years old.

2.2.1 Bones and Muscles

Reiss and Tunnicliffe have also undertaken extensive research on children's ideas about the human body (Reiss & Tunnicliffe, 1999a, 2001; Reiss et al., 2002; Tunnicliffe & Reiss, 1999a). They have used a variety of approaches to establish children's ideas about the structure and the place of the different organs. As part of a larger study they wanted to find out what children from 5–11 years of age knew about the human skeleton. The children were asked to draw what they thought was inside their bodies and were given ten minutes to complete their drawings. Reiss and Tunnicliffe then made a ranking scale of seven levels, each reflecting different levels of biological understanding about the human skeleton (Reiss & Tunnicliffe, 1999a). As expected, the older the children were, the higher the average score was. However, there was a considerable range within each age group, where one 10/11-year-old child would be scored at level 6 and another 10/11-year-old would be scored at Level 3. The study also showed that a third of the youngest children 5/6 years of age had very little or no knowledge about bones (Reiss & Tunnicliffe, 1999a). Reiss and Tunnicliffe also designed a similar seven level scale for different levels of biological understanding about organs and organ systems (Reiss & Tunnicliffe, 2001; Tunnicliffe & Reiss, 1999a). Older children also scored higher here on average, as again expected. In the light of the results of their studies Reiss and Tunnicliffe say that they "believe that the children learn about organs as units and gradually piece them together so that eventually they form a mental model of a system, be it the skeleton, the digestive system, or gaseous exchange system" (Tunnicliffe & Reiss, 1999a, p. 32). Results from their studies also show that by the time the pupils they studied were eight years old they mostly had a broad knowledge of the internal structure of the body and were aware of a wide variety of organs although they did not know how the organs were connected or how they were part of an organ system (Reiss & Tunnicliffe, 2001).

In the SPACE research the children tended to draw bones that they could feel so they were aware that there were bones in arms and legs and many also knew about ribs and the skull, but they were usually not aware of, or not sure about, the pelvis or the spine. However, 45% of upper juniors (10/11 years old) drew no bones in their bodies when asked to draw what they thought was inside their bodies. When asked about their muscles most of the children thought of their arms and possibly legs, but they did not seem to realize that all movement of the body involve muscles. The next most often mentioned location was fingers but this was only mentioned by primary, 5–7 years old (Osborne et al., 1992).

Surprising results emerged from other research that looked at children's understanding about the human body but where children were asked to draw their skeleton inside a "silhouette." Many children under six years old drew a bag of bones or little bones just all around the body even if they seemed aware of the fact that the skeleton holds the body up. Drawings from 6–8-year-old children were more advanced on the whole, but none of them showed long bones stretching from one articulation to another or in any coherent position in relation to muscle movement (Guichard, 1995). Another study of 112 children who had never had any formal teaching about the skeleton showed that 97% of children under 8 years old do not see the skeleton as a functional structure: They represented the skeleton as "a bag of bones," a chain of knuckle-bones, fish knuckle-bones or "stick figures" (Guichard, 1995, p. 248). In order to explore how 6–8-year-old children imagine muscles, it was explained to them in Guichard's study what muscles actually were and how they were positioned. In 86% of cases the children subsequently drew muscles either attached to something between bones or as a layer under the skin. In the cases when a muscle was drawn linked to a bone, its position would not enable it to initiate movement, which indicates that the children have no concept of the role of muscles in movement (Guichard, 1995).

––––––

2.2.2 Heart, Blood Circulation, and Lungs

Young children (Age 5–7) know more or less where the heart is located in the body and children of all ages give it a Valentine shape (Carey, 1985;

Osborne et al., 1992). According to Carey (1985) the heart is generally the first internal organ that children know about, although children under 9 years old knew little about its function. They rather described its psychological or social function or how it behaves, like: "It ticks."; "It pumps when you run fast."; "You can hear it beat."; and "The heart is essential for life." Then there were comments like: "The heart makes blood."; and "Blood comes from the heart."; but without any explanations about blood circulation. The internal organs are assigned a static function, the heart is for love, the lungs for breathing, the stomach for eating, the brain for thinking (Carey, 1985). In the SPACE study the younger children, 5–7 years old, typically knew about the heartbeat with a greater depth of biological knowledge shown among older children: The majority of them know that the heart pumps blood, though very few of them associated its function with gaseous exchange (Osborne et al., 1992). Ideas of the purpose of blood and how it is carried around the body were complex and mixed. Both primary (Age 5–7) and upper juniors (Age 10–11) described blood as being necessary to keep you alive, but a number of younger children tended to think that blood moved just by itself. Lower juniors (Age 8–9), however, surprisingly, compared to upper juniors, showed a level of understanding that indicated some knowledge of the circulatory process by saying that blood ran or moved through the body and many lower and upper juniors mentioned veins (Osborne et al., 1992).

The SPACE research also attempted to gather information about children's understanding of gaseous exchange. The ideas fell into three categories. First, were ideas of everyday nature like the air we breathe goes into our tummy, keeps us alive, and well. In the second category some greater knowledge was revealed about the organs' role in respiration and there was an improved level of understanding. In the final category were ideas that were very rare but showed knowledge of gaseous exchange, and there was a high correlation between these children and those who drew the lungs on their drawings of what is inside your body (Osborne et al., 1992).

2.2.3 Digestion

A number of studies have examined children's ideas about digestion. Young children (preschool children) seem to relate the stomach to breathing, blood, energy, and strength. By the age of seven they start realizing that the stomach helps to break up or digest food and later they understand that food is transferred elsewhere after being in the stomach (Carey, 1985; Driver et al., 1994). Young children are aware that the body changes as you grow and that not eating leads to the body becoming thin (Rowlands, 2001).

That food undergoes a process of transformation in the stomach is not generally known until Age 11 according to Carey (1985). Younger children thought that food went into various parts of the body or that food was discharged. The digestive process was understood as chewing and making food smaller and smaller. When asked about what happened to the food you eat, most 5-year-olds knew that it went to the stomach but thought that it stayed there unchanged. The few that knew that it was transported to other parts of the body imagined that it was transported unchanged. By the age of 6 almost all children knew that food makes its way to all parts of the body but only very few of children aged 9–11 "knew that food is changed in the stomach and brings about its desired effects by being broken down into altered substances that are carried to tissues throughout the body" (p. 45). Young children (Age 5–7) relate food and eating to growing, being healthy, being strong, and not dying—that is, their explanations are psychological, because they are explained in terms of beliefs (Carey, 1985). According to Driver et al. (1994), different studies show that when children draw the stomach, they draw it usually bigger and put it lower than it really is and the youngest children appear to relate the stomach to strength, energy, and also to breathing (Driver et al., 1994). In the SPACE study the children were asked to show on a drawing what happened to food in their body. At the simplest level children would simply draw a big bag containing untransformed food and no tube whatsoever, or food that was distributed all through the body. However, the youngest children produced drawings that show a lack of recognition of connection between the mouth and the stomach, and some children who drew food in all parts of the body would qualify this by saying that the food went into the blood and the blood goes elsewhere. These sort of statements suggest that they recognize that there is a process of some kind of transformation but their drawings may show the limits of their ability to represent food (Osborne et al., 1992).

The next feature to emerge in children's responses was to draw two tubes from the mouth to the stomach: one tube for drink and one for solids. An explanation given by teachers for this was that children hear comments such as "it went the wrong way," as used in everyday language (in Icelandic too!). When children drew only one tube from mouth to stomach, however, progression was seen towards scientific understanding although such drawings show a lack of understanding of what happens beyond the stomach, but this seems to be the hardest aspect for most children to understand. No primary (5–7 years old) indicated any signs of the digestive tract leading from the stomach and only the minority of the older children (8–11) did so. This indicates that excretion is possibly a relatively poorly understood process by children under 11 years old (Osborne et al., 1992).

However some of the older children made drawings that showed the digestive tract and located the stomach and the rest of the digestive system in approximately correct position, which supports the view that children's biological knowledge develops substantially between the ages of 5 to 10 (Osborne et al., 1992).

In a study on children's conceptions on the structure and function of the digestive system (Teixeira, 2000) children of Age 4, 6, 8, and 10 were interviewed. Each child was given a bar of chocolate and asked to eat it and draw the way the chocolate/food passes through the body. The children had not had any formal instruction about digestion. The results indicate that children possess biological knowledge as an independent knowledge domain from the age of 4 and that there are gradual changes in children's ideas, so that by the age of 10 they explain the function of the digestive system in terms related to the function of the organs, that is, as biological explanations. Taxeira also suggests that children's theory is built on the application of empirical knowledge and her findings indicate that biological knowledge is constructed from inferences drawn from daily experience and empirical knowledge is applied to related contexts (Hatano & Inagaki, 1997; Teixeira, 2000).

A study of children aged 4–8 on thinking about digestion and respiration (Toyama, 2000) showed that the children of all ages knew that a lack of food and breathing can damage our body. The children also knew that food changes inside the body, but they seldom referred to biological transformation. They also assumed that air warms the inside of the body. Preschoolers knew that food is necessary for your health and growth and the older children consistently understood biological transformation of food. The study suggests that by the age of 4 or 5 children seem to have a sufficient insight for accepting some material transformation of food in which "food goes to various parts of the body and turns into our bodies" (p. 229). Therefore, these results and the results of other studies mentioned here suggest that biological insights about transformation of food are accepted by children earlier than previously was claimed, for example by Carey (1985).

A Portuguese study about children's conceptions about digestion showed that most children in year 1 and 2 of the Primary school represented a body structure composed of a month linked to a "neck sac." These children had not had any formal instruction about the digestive system. Some children drew a continuous tube. Others drew a sack not linked to the mouth or drew food free in the body. Their drawings did not in general show well organized body structures but when asked about the organs they knew, they mentioned: mouth, gullet, stomach and gut, bladder, heart, or lungs (Carvalho, Silva, Lima, & Coquet, 2004). These children were also given something to eat, as in Teixeira's study (2000), but here they got a cookie. The majority of the drawings of year 3 and 4 pupils who all had had formal teaching about the digestive system showed great confusion about what lies beneath the stomach. However, all the year 3 pupils that were asked to write down a short text about the digestive system mentioned the correct sequence, suggesting that they learned it by heart (Carvalho et al., 2004). A high percentage of year 1 and 2 pupils represented the entire cookie inside the body, but none of the pupils in year 3 and 4 did. In year 1, 2, and 3 the representation of little pieces is most significant. From interviews with year 1 pupils undertaken in order to understand the meaning of their drawings, it became clear that they had different ideas regarding the cookie digestion, including everything from entire cookie, little pieces, and dissolved material passing into the body. They knew that the cookie went through the entire body even though they did not know the exact way and they were unable to draw the digestive system. This was also the case in Teixeira's (2000) study of digestion. Although the mechanical and chemical processes of digestion are not generally emphasized in Portuguese school teaching, the change or transformation of a cookie or other food into a creamy mixture in the mouth is an ordinary daily sensation. Therefore it is surprising to see that a number of year 1 and 2 children respectively drew the entire cookie inside their body as if they had swallowed it in one piece without chewing it. According to Carvalho et al., it is possible that this kind of schematisation aims to represent the cookie symbolically rather than realistically. Then such a drawing may then signify that "it is a cookie that I just ate" (Carvalho et al., 2004).

In an investigation of young children's understandings of excretion and the digestive system, children in years 3 and 5 (7/8 and 9/10 years old) were asked to draw the pathway from the mouth and through the body and out. The drawings were analyzed by a six point scale which Tunnicliffe devised in the light of three stages of understanding suggested by Clément (2003). The results suggest that both age groups lack an understanding of the structure of the excretory system but have a better understanding of the digestive system where they know that there is a continuous tube from mouth to anus. Excretion is not really mentioned in the British National Curriculum and usually the body systems are taught in isolation so the children are not offered "a holistic integrated view" (Tunnicliffe, 2004, p. 10).

2.2.4 Brain

There are very few studies that discuss children's ideas about the brain. Ideas about the brain are mainly mentioned when talking about children's ideas generally. According to Carey (1985), young children see the brain as the place where mental activity takes place and say the brain is for thinking. In the SPACE study, the brain is said to be one of the main organs named by the children, along with the heart, stomach, and bones (Osborne et al., 1992). This is also so in a case study about the human body where the brain was among the organs the children mentioned and drew (Frost, 1997). In a study of 8–9-year-old children, undertaken by Cuthbert (2000), the majority drew brains when asked to draw what was inside the body.

2.2.5 Reproduction

Carey (1985) is one of the few studies to discuss children's ideas about reproduction. She also discusses the results of two older studies: Bernstein and Cowan (1975) and Goldman and Goldman (1982). According to Carey both studies were extensive and both involved children ranging widely in age, from 3 to 12 in the former case and from 5 to 16 in the latter. Both studies probed the origin of babies and the role of the mother and father in producing babies and both found the same developmental progression (Carey, 1985). The youngest children (3–4 and even 5 years old) presuppose that babies have always existed. Before birth they just lived somewhere else, in their mother's tummy or in a store. The child interprets the question "How do people get babies?" as "Where do babies come from?"—i.e., as a problem to discover where the baby was before it was present in the family but not how it came into existence. When the youngest children were asked "How are babies made?"; their answers were profoundly nonbiological and the children focused on the word "made" and described it as a process of manufacture. So preschool children see the origins of babies only in terms of intentional behavior of the parents who go to the shop to buy them or who make them and place them in the mother's tummy. These children knew nothing about how a person's body produces a baby (Carey, 1985).

According to Carey, young children construct theories from their own experiences and understandings. In the Bernstein and Cowan study (1975) a child explained that to create a baby parents would first have to buy a duck, which turns into a rabbit, which then turns into a baby. The child had learned this from a book which tried to explain the issue of the mechanics of sex by starting with ducks and rabbits. Primary school children, however, know more about the facts of life and may well mention eggs and sperms, "but their understanding of the process of birth is based on analogies to activities for which physiological mechanisms are not yet known" (Carey, p. 58). By the age of 10 the children understand that bodily processes result in a new baby (Carey, 1985).

Reiss and Tunnicliffe's study (2001) of students' understanding of their internal structure showed that none of the children in Reception year and Year 2, drew or labeled any reproductive organs when asked to draw what they thought was inside themselves and only two children in Year 3 (both boys) drew or labeled any reproductive organs. However, half of the girls in Year 6 rather drew male reproductive organs than female and this was also the case for the Year 9 girls—they were more likely to draw male production organs than were male students of the same age.

This review of the literature about children's ideas about the body suggests that young school children (Age 4–7) know the names of the body parts that they can see and touch and they know bones and other organs that are more easily sensed than others, like the heart which is the most commonly known organ although most children represent it like a V-shaped Valentine's heart. Children at this age (4–7) know that we need food to live and that the food goes from the mouth and into the stomach and that it changes in the body but their understanding of what happens in and beyond the stomach is very vague. Their understanding of processes such as the blood circulation process and the digestive process is also very vague. The studies suggest, however, that biological knowledge develops extensively between the age of 4 and 9 although it is not until the age of 10/11 that children seem to understand the main processes like the blood circulation process and the digestive process.

2.3 Interaction—Learning From Each Other and From the Teacher

The literature about social interaction in the classroom builds a lot on the views expressed earlier, that conversations between children and adults are crucial for cognitive development. A case study undertaken by Appleton & Asoko (1996) of a teacher's progress towards using a constructivist view of learning highlighted some of the problems of translating constructivist teaching with extensive cooperative interaction into classroom practice. The teacher, who was an expert classroom teacher, was faced with teaching science in a normal primary teaching situation. Results from the study suggest that finding out what ideas the children held tended not to be used as a basis for planning a teaching strategy to lead to conceptual development. The children's ideas were instead used to decide which activities the whole class would be involved in. The researchers emphasize the need to articulate clearly constructivist principles and how to implement constructivist ideas in the classroom in in-service programs and in teacher education (Appleton & Asoko, 1996).

Hedegaard (1999) builds on Vygotsky's theory in her own research and argues that a child's conceptual development is shaped by the social practice of the institution which the child attends. Through participation in the social practice of the school the child conceives the content and methods that characterize the social practice of everyday activities in the classroom as well as subject matters activities. According to her, the conceived knowledge and methods of school are transformed into individual knowledge and thinking by the child's active use of the formal subject matter content and methods in motivated class activity where children cooperate and communicate with each other (Hedegaard, 1999).

Collaborative interaction in the classroom involves the mutual construction of shared meaning and it requires both communication skills and an understanding of others' minds (Ogden, 2000). Collaborative or cooperative learning means learning or working together, and during and through cooperative interaction it is assumed that children have opportunities for

> planning strategies, verifying ideas and encountering the symbolic representation of intellectual acts through peer communication. The ineffectiveness of an individual in a cooperative learning situation may reflect failure to distribute limited cognitive resources adequately between the various complex components of the task. (Topping, 1992, p. 153)

Every child is different and every child has its own store of experiences, ideas, and knowledge that he or she uses to make sense of the world, which is important for interaction in the classroom because most learning takes place in a social context (Fisher, 2005) or as Bruner puts it: "Making sense is a social process" (Bruner & Haste, 1987). It is through "knowledgeable others" that a child's potential for learning is revealed and these can be anybody: parents, teachers, or other adults but also siblings, friends, and peers (Bruner & Haste, 1987; Fisher, 2005). According to Fisher (2005), "children learn best when they have access to the generative power of those around them" (p. 92), and the basis of "success-through-others" is communication. The best conditions for learning exist when children have a challenge that extends their cognition, and therefore the role of the teacher is to provide the social and cognitive framework for learning. Co-operative learning can be undertaken through discussion and problem-solving or through production tasks, and it can be in the form of peer tutoring, small group-work or working in large groups (Fisher, 2005). Peer tutoring is a process where all involved can benefit: the tutor, the child that is being helped, and the teacher. The tutor (helping the child) can benefit from the helping role; for example, from the fulfilment of being able to help someone and by consolidating their knowledge, filling in gaps and extending their own conceptual frameworks.

The child being helped can benefit greatly as it gets extra individual attention and, if it works well, the spoken interaction with a peer is of a personal and powerful kind. Peer tutoring can free teachers from some of the routine work involved in monitoring a whole class (Fisher, 2005).

In a study on collaborative tasks, children were observed participating in paired activities in which they were required to work with one other child. Reception and year one children were observed engaging to some extent in reciprocal interaction (Ogden, 2000). According to Ogden, previous shared experiences facilitate children in getting to know and understand others' perspectives:

> ...it may follow that forming working partnerships with classmates for particular educational activities may enable them to share meanings through repeated joint involvement, thus allowing greater collaborative and reciprocal interaction and the extension of shared meanings. (p. 224)

Learning in a group can cover a range of possibilities. The children can be working together in a group setting but on individual tasks. When children are sitting in groups, but are not working together as a group, this can distract children from their work. Then they would work better on individual or paired tasks and sit individually or in pairs. However, children need to be in groups for cooperative group work (Fisher, 2005). But it may be important how the group operates. According to Fisher (2005) there are four kinds of learning in groups:

1. Children working together on individual tasks sitting in groups;
2. Children working in a group on shared tasks with a joint outcome, for example a problem-solving or construction task;
3. Children working together on activities which contribute to a joint outcome (e.g., research task or a story); and
4. Children working in a group with an adult who guides the learning of a small group through different planned tasks (p. 98).

The difference between 1 and 2 is not obvious but manifests that working in groups, which on the surface seems similar, can be of many kinds.

There are also many benefits that can be gained from learning in large groups, which is the traditional approach of organizing and teaching children in classes. A class of children is a community and should provide the benefits of a community such as support and resources and as such can provide the essential structure, purpose, and control needed for successful learning. The stimulus to challenge and extend the child involves the use of questioning, explanations, and statements that aim to involve all the pupils in the class in active thinking and responding, which again correlates with higher level of pupil performance (Fisher, 2005).

Joint classroom experience means that children have acquired a common knowledge with their peers that may encourage responsive communication, listening to others, and sensitivity to the views of others. Through engagement in social action with peers or classmates, children are presented with points of view or ideas that differ from their own in response to which their existing ideas may develop and change (Ogden, 2000). Therefore the classroom experience may provide them not only with opportunities for collaboration, but also the development of understanding of working in task settings, because through social interaction children gain knowledge, skills, and understanding of social practices. Ogden's work emphasizes the need to create learning environments that encourage the exchange of ideas and actions. It also highlights the value of early interactive experiences for peer work, where children can experience the perspectives of others.

Reiss and Tunnicliffe (1999b) discuss the considerable influence the learning environment has on the development of ideas. They emphasize that for most people learning "takes place principally when two conditions are satisfied: first, when what is being learned is of a personal meaning to the learner; secondly when there is a social environment for learning" (p. 14). The teacher has to bear this in mind and has to provide learning opportunities that include a combination of both.

Since computers are now based in most primary classrooms there are a number of studies that discuss young children's collaborative interactions when working together on the computer. Most of these studies are supported by both Piaget's and Vygotsky's theories of development (Dalgarno, 2001; Hyun, 2005; Lomangino, Nicholson, & Sulzby, 1999; Shahrimin & Butterworth, 2002).

Dalgarno (2001) looked at the consequences of constructivist theories of teaching and learning for computer-based learning. He suggests that although dialectical constructivism emphasizes the role of social interaction and support or scaffolding for learners that would normally be provided either by the teacher or the peers, computer software tools could in some cases also fulfil this role.

In a study by Hyun (2005) that explored characteristics of 5- and 6-year-old children's peer dynamics in a computer-based classroom in the United States, paired children who differed in computer proficiencies but shared similar interests were found to work very well. The research exemplifies Vygotsky's dialectical constructivist perspective on peer teaching and learning characteristics. "Their conversations displayed self-confidence, multiple perspective-taking skills, and reflective self-assessment" (p. 69). More capable children, however, were not always sufficiently challenged intellectually by peers that had capabilities that were still in an emerging stage.

Collaborative computer use is often associated with the social nature of interaction. A study that investigated how interactive patterns develop in collaborative activity while working on a computer showed that even with minimal adult involvement, children exhibit many constructive patterns of interaction while working collaboratively on the computer and they rely on each other as resources. With their collaboration the children did provide each other with scaffolding that is considered important for development, although children's support often consisted of direction rather than teaching (Lomangino et al., 1999). Both in this particular study and in a study by Mercer, Wegerif, and Dawes (1999) disagreement and disputational talk sometimes occurred between the children who were working together and these were rarely followed with justification to help peers understand the reason for the opposition, which means the teacher has to have in mind showing the children how to focus on evaluating the product rather than the person (Lomangino et al., 1999).

A study that looked at teachers' use of talk in a whole class teaching context examined whether teachers make meaningful connections for children, in whole class teaching, between their past and current learning experiences (Myhill & Brackley, 2004). The learner actively constructs knowledge and understanding through interaction between prior knowledge and new knowledge and therefore classroom interaction should be established which builds on prior knowledge in order to help pupils understand underlying principles rather than having a superficial understanding of specific experiences. However, where the talk in whole class discussion is dominated by a minority of children, it will not give teachers the opportunity to assess accurately children's prior knowledge (Myhill & Brackley, 2004). Research shows that teacher-led questioning and explanation still dominates in primary classrooms and so do teacher's objectives while research into the particular nature of classroom discourse has shown that it is dominated by questions and statements from the teacher (Burns & Myhill, 2004; Myhill, 2003; Myhill & Brackley, 2004). According to Burns and Myhill (2004) it seems that pupil participation is differently experienced within the class with more active engagement by higher achievers than low achievers. Teachers in their study rewarded those who were compliant by inviting responses from those who had put up their hands for answering. Their results also showed quantitative dominance of the talk by the teacher, with a discourse pattern of teacher–pupil–teacher–pupil being the most common with little feedback from the teacher. Classroom discourse is, however, often described in terms of Initiation–Response–Feedback. The questions identified in their study "were factual, speculative, procedural, and process. Statement forms were informing, explaining, instructing, socialising, and elaborating" (p. 45). The researches argue that this kind of discourse situation does not support or scaffold pupils' learning, because the interaction is largely a case of pupils participating on request and not interactive in the sense of engaging children in more active talk to develop and create their understanding:

> It is with the *interplay* between the pupils' talk and their learning needs and the teachers' use of the differing forms and functions of language to enable children to think and explore their learning through a *real dialogue* that teachers should be concerned. (Burns & Myhill, 2004, p. 48)

Once the teachers in Myhill's study (2003) had elicited the correct answer in the dialogue process they moved onto further examples and repeated the sequence. According to the study the emphasis in the lessons is upon teaching and there is little evidence in the lesson sequence of how the children will acquire an understanding:

> The teacher is able to talk confidently about the objectives of the teaching sequence and the concepts she wants the children to learn, but she does not reflect upon how her own teaching decisions relate to children's learning. (p. 363)

Myhill emphasizes that developing effective learning demands teaching contexts which provide space for children's thinking and talking:

> Both listening and speaking are potentially active cognitive processes so it is important to emphasize teacher talk and listening, as well as pupil talk and listening in order to establish cooperative construction of meaning. (p. 368)

Establishing what children already know creates an opportunity to clarify misconceptions that might occur, to build on shared experiences, and to act on unanticipated understandings that might become apparent (Myhill & Brackley, 2004). Feedback on performance and in discussion is therefore a powerful tool in assisting learning. Children need feedback on their past efforts, on their ideas, and on their actions to help them identify what will lead to future success and further praise (Fisher, 2005). Feedback from the teacher has an important role to play in children's learning and also on their view of themselves as learners. Feedback from previous success or failure has influence on children's future motivation for learning (Harlen, Marco, Reed, & Schilling, 2003).

Mime (facial expressions) can also be seen as a feedback and can be very powerful. The use of mime to exemplify the active and passive was a conscious choice of strategy in the teacher observed by Myhill (2003):

> ... her reflections after the lesson record that the mime was a motivational tool, a tactic to *"help keep children on task."* She tries to *"present the information in as many different ways as possible, visual and verbal"* and hopes the mime would *"reinforce the teaching point."* (p. 362)

The words "enthusiasm" and "inspiring" describe good teaching and learning. According to Shakespeare (2004) these words are not enough because "if teachers cannot communicate, they cannot teach." To get a "Wow" from a group of pupils when demonstrating, the teacher has to engage emotions such as curiosity and excitement. Shakespeare asks questions where he encourages teachers to think about how they draw pupils' attention: what words they use, how they say them, and where the pupils should be located in the classroom; the teachers' actions and the way he or she carries them out throughout demonstrations; and how teachers go about planning to convey and generate interest, with individuals, small groups, and a whole class.

In a study that focuses on learning from human tutoring, tutoring is seen as existing in three different forms: tutor-centered, student-centered and interactive (Chi, Siler, Jeong, Yamauchi, & Hausmann, 2001):

> Tutoring skills refer to the pedagogical skills of knowing when to give feedback, scaffoldings, and explanations, when to hold back error corrections and allow students to infer that an error has been made, and so forth. (p. 472)

According to Graesser, Person, and Magliano (1995), a typical tutoring process has five broad steps that they call the "tutoring frame":

1. Tutor asks an initiating question
2. Students provide preliminary answers
3. Tutor gives feedback on whether answer is correct or not
4. Tutor scaffolds to improve student's knowledge and understanding
5. Tutor values/estimates student's understanding of the answer (Graesser et al., 1995).

In a classroom setting the teacher may start with an initiating question, the students give some answers, the teacher comments or gives feedback on the answer, and then the dialogue pattern of a classroom typically stops here (Burns & Myhill, 2004; Chi et al., 2001). However, in tutoring, the two additional steps are taken. In the fourth step the tutor continues with a "scaffolding episode," like that described by Vygotsky and Bruner where the adult or someone more capable "guides the child to develop and achieve to the child's fullest potential" (Chi et al., 2001, p. 473). The fifth step, according to the tutoring frame, is for the tutor to check and evaluate students' understanding by asking questions in order also to ask the students to evaluate their own understanding (Chi et al., 2001). These authors suggest in the light of their research that when tutors have the role of guiding students in a way that does not necessarily give direct correct feedback but is more open-ended and virtually never provides answers to problems, the students are more motivated and curious. If tutors ask questions in order to enable students to correct their errors and line of reasoning they give students more opportunities to construct knowledge. However, learning seems to occur not only through participation in a dialogue. Some learning also occurs through observing and overhearing others, although a study on students' performance on a problem-solving task when one group did the task and participated in discussion with the tutor while another group overheard the dialogues of the participants showed that participants performed better than those that overheard, suggesting that interaction facilitates performance (Chi et al., 2001). Chi and colleagues found that students learn just as well without the benefit of hearing any tutor explanation or feedback. If explanations and feedback are used, however, students learn more if the explanations provide some integration, contain relevant information, repeat the information and are rephrased in everyday language. A more interactive style of tutoring can be beneficial and motivating, leading to a greater enjoyment of learning. Results show that students' construction from interaction is important for learning suggesting that interactive tutoring ought to implement ways to elicit constructive students' responses and that one way to elicit construction is through scaffolding:

> The important thing is to determine what kind of scaffolding prompts to give and when to give them and to bear in mind that natural language understanding and an understanding of the content domain are crucial for appropriate scaffolding. (p. 518)

According to Harlen (2004a) much of teachers' talk in the classroom is in the form of questions, but there are many forms of questions and many different purposes in asking them. Harlen says that what we describe and discuss in the classroom is often more a question-and-answer session, where the teacher does most of the questioning. Such exchanges lack the richness of genuine dialogue because the value of real dialogue lies in the learners hearing the ideas expressed by others and in having to share their own

(Harlen, 2004a). In science learning we want children to link new experience to prior experience in order to develop further and bigger ideas. We help children test existing ideas. If through their investigations they find that the evidence supports the prediction, the idea is developed further (Harlen, 2004b). A learner is a part of a group and he or she contributes to the group and gains from it. This fits the idea of "social constructivism" that suggests that understanding exists in social interaction when learners as part of a group or a class discuss and attempt to give explanations and listen to others justifying their explanations (Harlen, 2004b).

Mortimer and Scott (2003) distinguish between "authoritative" and "dialogic" talk, where the former is controlled by the teacher but the later is controlled by the teacher together with the children and the direction of the talk depends on what the children say, and not just on what the teacher says (Appleton & Asoko, 1996; Mortimer & Scott, 2003). Both types have their place in the classroom, but clearly dialogic talk gives more opportunity for children to express their ideas and develop their understanding by making links to their current knowledge (Harlen, 2004b).

Mortimer and Scott (2003) talk about the different patterns of discourse. According to them the common pattern in classrooms is the "patterns of three": teacher–student–teacher with response, feedback, or evaluation from the teacher. They however recommend the pattern involving feedback rather than evaluation even though they do not distinguish it very clearly. The pattern of discourse can also be a chain of interaction, with a longer sequence, where the elaborative feedback from the teacher is followed by further responses by the teacher and so on. Mortimer and Scott also discuss the idea of the interactive–non-interactive dimension where talk can be interactive in the sense of allowing people to participate or non-interactive in the sense of excluding people from participation. By combining the two dimensions any sequence of the talk that takes place in the classroom can be located on a continuum between interactive and non-interactive, on the one hand, and between dialogic and authoritative talk, on the other. In the interactive/authoritative communicative approach, the teacher can maintain a great deal of interaction with their students, but do not necessarily pay attention to the students' ideas and is looking for only one answer. The interactive/dialogic communicative approach is different from the authoritative in that here the teacher listens to and takes account of students' points of view, even though their ideas might be different from the scientific view. Dialogic interactions often occur when the teacher tries to elicit students' views. The best example of the non-interactive/authoritative approach in action is probably the formal lecture while the non-interactive/dialogic communicative approach is when a teacher makes a statement that addresses the students' points of view, but at the same time does not call for any interaction with the students. These approaches of communication provide, according to Mortimer and Scott, a very useful tool for identifying the different ways in which teachers can work with their students in developing ideas:

> What this aspect of analysis does *not* tell us is anything about the way in which each communicative approach is actually achieved in the classroom, through the particular patterns of discourse and forms of intervention used by the teacher. (Mortimer & Scott, 2003, p. 40)

According to Elstgeest (1985b) the type of questions asked by the teacher is very important:

> A good question is a stimulating question which is an invitation to a closer look, a new experiment or a fresh exercise. The right question leads to where the answer can be found: to the real objects or events under study, there where the solution lies hidden. The right question asks the children to show rather than to say the answer. (p. 37)

Elstgeest calls these kinds of questions "productive" questions, because they stimulate productive activity, but according to Harlen (1993) there are three aspects of questioning to be considered and these are: form, timing, and content. Form refers to the way in which the question is expressed, whether open or closed and person-centered or subject-centered. Timing is important: teachers often ask questions at the wrong time because they are eager to press ahead too quickly so it is important to keep the timing of a question in mind. The content of a question should depend on where we are in the discussion. At the start of a project the purpose of the question is usually to find out the ideas that children have about the subject matter (Harlen, 1993). Discussion can also take part during the lesson and will often be a part of practical work. When the teacher is involved in the discussion the purpose may be to monitor progress, to offer suggestions, to clarify, to encourage exchange of views, and to assess (Harlen & Qualter, 2004).

Elstgeest (1985a) put forward some guiding points for planning discussion in the primary classroom. One of the points he made is that the teacher has to appreciate children's efforts whatever their results may be. He or she also has to avoid making suggestions that the only worthwhile result is the "right answer" and has to be ready to accept what the children find from their investigations as the "right answer" providing they have themselves used their evidence and their reasoning. The last point made by Elstgeest is that the teacher has to allow the children to make sense of their observations for themselves without imposing explanations that are not within their experience and comprehension.

In this review of the literature on interaction in the classroom I have not been able to find many references where children, on their own initiative, take a real lead in teaching their peers. This is an important area that needs to be explored. In only one case study (Frost, 1997) where children in year 2 (Age 6–7) were working on a topic about the body, "information from the children" was listed as how the section about the lungs was taught. One child in the class had become interested in finding out about the lungs, discussed it with his parents, and in class described the location and the function of the lungs.

Thus, the importance of peer interaction is emphasized in the literature and is thought to be especially fruitful when working on computer tasks. However, there is relatively little discussion of the effects of peer interaction in relation to Vygotsky's ideas about the ZPD as discussed by Carugati (1999) and Kutnick (2001) in Section 2.1, but according to them, cooperative or symmetrical relationships should be taken into account as a site for ZPD.

2.4   Teaching About the Body

In the SPACE study (Osborne et al., 1992) the aim was not only to provide some insight into children's existing knowledge and understanding but also to devise some intervention activities that could develop children's thinking and knowledge. The intervention phase was designed with a range of activities to provide an opportunity for children to represent and clarify their thinking, and this was generally done through group discussion and drawings. The criterion for selection of activities was that they should "require the *active processing* of information" and they were

> also designed to broaden children's schematic knowledge, extend their vo-
> cabulary and where appropriate, generate conflict between their thinking
> and experience which would lead to a re-evaluation of their ideas. (p. 46)

The intervention activities included sorting activities, discussion activities, modeling, or making activities and investigations. In the classrooms a lot of secondary sources and equipment were set out for further information and investigations. The analysis of the SPACE data shows two contrasting effects of the intervention. One is that there are many instances to show that the knowledge and understanding of the primary (Age 5–7) have improved; and the other is that the intervention failed to make any significant effect with lower (Age 8–9) and upper juniors (Age 10–11).

The results showed that before the intervention the majority of children perceived muscles as being in arms and legs and only a small minority indicated that muscles are in other parts of the body.

The intervention improved awareness of muscles in other parts of the body and a significant increase in the number of the youngest children in the study that stated that muscles were to be found everywhere in the body. There was also a significant change after the intervention in the number of infants who indicated the presence of the brain and lungs. Both of these increased after intervention. Prior to the intervention infants differed significantly from upper and lower juniors but after the intervention they did not. The understanding of the location of the heart also improved as a result of the intervention.

The most significant change was the increase in the average number of organs or parts of the body drawn by each child. The number of organs drawn by lower juniors also improved, but the upper juniors showed no improvement, which suggests that they knew of them already.

When asked to explain what happened to food and drink inside the body, more of the youngest children in the study showed the inclusion of a tube (or tubes) between the mouth and "tummy," while there were no significant changes for the older children. However, there was a significant increase in the number of lower juniors who said that the heart was responsible for circulating blood when asked about the function and purpose of blood. Upper juniors showed a significant decrease in the number who showed the heart as a Valentine-shaped object and also a decrease in explanations in terms of the fact that the function of blood is to keep us alive: instead there was an increase in the number who explained that blood runs around the body. These results suggest that such an invention had positive effects for the infant children but very little effect for older children who already may have a reasonable level of knowledge in the domain. The differences with age existed prior to the intervention, indicating the development in children's biological knowledge during the transition from Age 5/6 to Age 8/9.

Identical pretests and posttests were carried out in Guichard's (1995) study where children were asked to draw a skeleton inside a "silhouette." The pretest was carried out before the children had been introduced to the subject and the posttest was carried out after either a traditional teaching session or exhibits that involved various exercises aimed at improving knowledge and understanding. These exhibits were designed from the framework of a child-oriented interactive exhibition in a science museum in France and as learning tools were based on the theory of formative evaluation and the understanding of learners' existing ideas. The study was carried out in classrooms and in children's interactive exhibition with exhibits and exercises designed to help children to understand the function of the skeleton and muscles. According to Guichard, the designer, like the teacher, must create an environment in which the exhibit has a meaning for the visitor or pupil who will decipher the exhibit according to his or her own framework. In the classroom setting, two primary school teachers with classes of six to seven years old, were asked to do a presentation of a skeleton in the class, using an elementary anatomical diagram. They gave the names of the principal bones and gave each child a photocopy of a skeleton and the children were asked to learn for the lesson the next day. The immediate results were promising. The following day the children were able to draw a rather precise and structured picture of the skeleton and name a few bones. However, six months later the children were asked to do the same tasks and results showed that "very few children had acquired a functional concept of the skeleton and its role in movement" (Guichard, 1995, p. 249). Most of the pupils drew skeletons according to their previous conceptions and a few of the elements emerging from the class lesson, but the general idea was still very vague. The ribs were still drawn as "fish-bones" as in the pretest and the technical terms of the skeleton were still very confused.

Results from the exhibits show that the exercises the children did had both short- and long-term effects on children's ideas of how their own skeleton works and almost all the children had acquired a solid understanding of the function of the bones. "Practical exercises for children therefore facilitate the acquisition of knowledge and an understanding of the role and functions of the human skeleton" (Guichard, 1995, p. 252). Results from the exhibits also show that when children made mistakes they tried to correct their mistakes. This was the case in an exercise about muscles where the children were made to think about muscle movement in relation to articulations and how the muscle system functions as a whole. This shows how children's misconceptions and mistakes can be used to design efficient exhibits and how their drawings and ideas can be used to design exhibits that in turn will help develop children's knowledge on a long-term basis. Similar exercises can be used to develop efficient teaching tools for use in schools.

Reiss and Tunnicliffe believe that children learn about the body as units which they gradually piece together so they eventually form a "mental model of a system, be it the skeleton, the digestive system, or gaseous exchange system" (Tunnicliffe & Reiss, 1999a, p. 32). When learning, for example, about the skeleton, children start with bones in general, then progress to bones in particular places and then continue to look at and understand bone units such as leg bones or ribs. Tunnicliffe and Reiss conclude that children should be taught about the skeleton in parts or stages and point out a range of teaching approaches that would help children get a holistic overview of the skeleton. One approach is using a jigsaw where the children put different bone shapes together in order to form a skeleton. Another approach is to encourage the children to compare their own skeleton to the skeletons of other vertebrates. Models of skeletons can also be explored and also x-rays and ultra-scans. A visit to a hospital can be educative if it can be arranged and books and videos can give rich and integrated accounts of skeletons (Reiss & Tunnicliffe, 1999a).

Reiss et al. (2002) suggest that science education builds upon and extends the knowledge that children bring to science classes and, as with the bones, it seems that children learn first that they have certain individual organs. Then they realize that these organs are situated in a special location and then they come to realize that some organs function together and are together in functional units (e.g., the oesophagus is joined to the stomach). In some cases children then learn that a number of organs function together in a whole organ system. Therefore, teaching about the body, rather than starting teaching about the whole organ system and then going into more detail, should start by teaching or exploring individual organs and then helping children learn that they function together and they all are part of a functional system (Reiss et al., 2002).

The perspective of obstacles to learning is sometimes seen as an important issue for teaching. Clément (2003) proposes three complementary categories of obstacles: epistemological, didactical, and psychological. Epistemological obstacles come from contradictions between everyday life and scientific knowledge. Didactical obstacles come from contradictions between previous teaching and scientific knowledge. Psychological obstacles are however identified as "the self skin" and an example of a psychological obstacle is: "Our skin is also a permeable wall, supporting our communication with the others..." (p. 97). The results of Clément's study show that most university students in his study had forgotten that ingested liquid goes from the intestine into the blood and into the whole body through the circulatory system although this information had been given to them several times in primary and secondary school. According to Clément this failure to remember can be the result of several didactical obstacles:

- Many of the earlier French primary school textbooks often draw "the way of the food" from mouth to anus, with precise times: a boy is eating an apple at midday; 1 minute after, the apple is in the stomach, and 14 hours later it reaches the anus! There is a big mistake: that is not the "way of food," but the way of what is not digested! In these drawings, the blood is absent.

- In secondary school, the chapters on digestion, circulation, excretion are separate, without any cross referencing question. Each chapter had several interesting messages: the permeability of the tube walls is only mentioned without much relevance.
- The teaching and evaluation of biology is often too academic, and would need more connections with everyday life situations (p. 96).

Clément's work suggests that identification of didactical, epistemological, and psychological obstacles underlying the main conceptions could help in proposing more effective pedagogical strategies.

In research undertaken by Slaughter and Lyons (2003) a sample of Australian children were taught about the human body and how it functions to maintain life. A control group received no teaching. Both groups were tested at the beginning and also at the end by an interview (pretest/posttest). The experimental group were taught about vital body parts and processes: blood, heart, lungs, brain, kidneys, the role of food, digestion, bones, and muscles. A poster was used that showed the outline of the human body and the major blood vessels and there were cardboard cutouts of the main organs that could be affixed with Velcro to the poster when each topic was covered. Small cards that contained important information about each organ were used to ensure standardized training to all the children. When all the topics were covered all the children were given an organ each and were told to come up one at a time, attach it to the poster and tell the other children about it and its function. After the poster was completed the children participated in a drawing exercise where they were given a paper with an outline of the human body and crayons. Each child was asked to draw an organ in the right place and tell the rest of the children what they could remember about that particular organ (Slaughter & Lyons, 2003). Results showed that all the children in the training group showed significant increases in their body concepts. They also increased their level of understanding for individual vital body parts, while those who did not receive training showed no increase in their knowledge of the human body's vital functions (Slaughter & Lyons, 2003). This is not very surprising but does suggest the importance of different teaching methods and exercises.

Formal teaching about human biological systems in the Portuguese primary school does not start until the first term of school year 3 (Age 8). In the study of Carvalho et al. (2004) about children's ideas of the digestion system, one classroom from each year (year 1 up to year 4) was selected randomly within the same primary school. At the beginning, the children were asked to make a drawing in order to obtain information about their conceptions of digestion, and individual interviews were carried out with pupils whose drawings were difficult to understand, while year 3 pupils were asked to write short texts about digestion as mentioned in Section 2.2. The aim of the research was to identify learning obstacles; two types of obstacles were differentiated on the basis of the origins of children's concepts. These are the ones gained from the daily life (epistemological obstacles) and those gained from formal learning activities (didactical obstacles), as also discussed by Clément (2003).

After teaching about digestion, the children's conceptions of anatomy of the digestive tract changed completely, the drawings were more realistic and less symbolic. The results from this study show clearly that school teaching about digestion and the digestive system can cause major changes in pupils' thinking. The existence of a digestive tube with its stomach and intestine became the majority view after teaching (in years 3 and 4): "It is scientific knowledge that has to be constructed: it does not emerge simply from children's intuition" (p. 1124). The drawings of year 3 and 4 pupils suggest that the children were especially influenced by the figure of the digestive process in the school textbook as they did not draw a clear continuous digestive tube from the stomach to the anus (Carvalho et al., 2004). After teaching about digestion the great majority of year 3 and 4 children's drawings reproduced the textbook schema with no representation of the continuity of the digestive tract and showing confusion in the anatomy and connections of the small and large intestine. None of the 120 pupils in the study drew the passage of digested products into the blood. Twenty three per cent of year 3 pupils, however, mentioned "blood absorption" when asked to write a short text about the cookie digestion, although they did not show it in their drawings. However, the researchers think these comments were made after learning by heart rather than from understanding, as most of the primary school textbooks mention "blood absorption" without an explanation of what this means.

Most of the year 3 and year 4 children in this study had difficulties in drawing connections between the stomach, small intestine, large intestine, and anus and those difficulties could be associated with the textbook picture used during teaching about the digestive system. In the picture a part of the liver hides the continuity between the oesophagus and the stomach. The picture does not show the connection between the stomach and the small intestine and it does not show either the small intestine track or the continuity of the small intestine with the large intestine. As a whole the picture in the textbook used does not demonstrate the appropriate notion of the continuity of the digestive tube, from the mouth down to the anus. Such a picture can generate an important didactical obstacle that prevents children from getting the proper notion of the continuous digestive tube.

To avoid such didactical obstacles, textbook pictures must be clearer and pictures more schematic. It can, however, be more difficult to overcome the main epistemological obstacle which is the difficulty of understanding the passage of digested food into the blood, through the intestine and capillary walls, and then throughout the whole body. Carvalho et al. have a solution to this:

- ▪ The modification of the syllabus, with more emphasis on the issue;
- ▪ The modification of the earliest images of the digestive tube in primary school, with a clear drawing of the passage of the digested food from the intestine into the blood and then throughout the body; and
- ▪ A motivation of the pedagogy in the classrooms, to focus the teaching and learning sequence in order to overcome this epistemological obstacle (p. 1128).

A possible solution is also to ask the children to work on their own drawings, to analyze them, discuss amongst themselves, and so improve their own conceptions (Carvalho et al., 2004). Tunnicliffe (2004) also suggests a teaching strategy that can be useful when teaching children about the human body:

- ▪ To find out at the beginning of the work what children already know from their everyday life and from previous work at school;
- ▪ To find out what "body" words the children know and can recognize and what meaning they ascribe to them;
- ▪ To ask children to do drawings and explain what they draw (p. 10).

A case study of a class of year 2 children (Age 6 and 7) in London focused on learning about the body (Frost, 1997). The topic of "Ourselves" was taught for six weeks, taking up most of the teaching time every day and also involving nearly all areas of the curriculum. The children had previously studied "Ourselves" in the reception class (Age 4/5) but had focused mainly on external features and senses. Many methods and approaches were used and a life-size paper model of the human body acted as a summary of the work done. Each new piece of information learned about the body was written in words on one side of the figure, and on the other side there were drawings of the different parts and organs, added systematically as the topic progressed. Two models of the skeleton had also been up in the classroom and one of them showed not only the bones but also many of the organs in the body.

Each new system or subtopic started with discussion or another exercise so that the teacher could find out what the children already knew. Then she tended to give them different activities that they could work on together in small groups, but extensive use of different sources of information enabled the children to work by themselves at different rates. Day-to-day activities gave the teacher information about the children's learning and there were a few activities especially planned to help the teacher get information about what the children knew and could do. The first was a brainstorming exercise, then a problem-solving exercise, and third a drawing task that the children did individually both before and after teaching. Then there was an interview with each child. This study shows that good assessment tools are usually also good learning activities and sometimes it is neither possible nor useful to distinguish between them.

The drawings the children did after the teaching about the body showed that the number of body parts had increased, their locations were better understood and the drawings were more detailed. According to Frost it is likely that the increase in knowledge was greater than shown because children often know more than they show on a drawing.

When working on the skeleton the children undertook many different activities; for example, they discovered their own bones by feeling them, finding the bends in their bodies to find the joints. When learning about the heart the children did running exercises and were asked about what happened to their pulse when they exercised and why this happened. Learning about the lungs was done mainly through the teacher talking to the class and showing them pictures and allowing them to look up information in books. A good example here is of a child who was not usually interested in science but was asked to look up information on the lungs. Two weeks later when the teacher was starting to talk about the lungs the child put up his hand and told the class masses about the lungs, where they were in the body, and how they functioned. So when describing how this section was taught "information from the children" must be on the list. It was interest that led a small group of children to explore the brain. The brain was not something the teacher had envisaged teachingbut its inclusion was instigated by the children. The children worked independently in a small group but the teacher had spent a little time talking about how the brain sends messages to make the rest of the body work. To help them understand this she got the children to give messages to her arm in order to pick up a pencil. They played this game several times and through the activity their understanding improved.

Although drawings are a common activity during science lessons in the primary school, as described here in some of the literature, curriculum developers and researchers have not given them much attention (Heyes, Symington, & Martin, 1994). While curriculum developers have acknowledged that teachers should involve children in drawing activities there has not been any comprehensive consideration of what purposes could be achieved by involving pupils in drawing and what teaching strategies are available to achieve those purposes, and there has been relatively little interest in research, which has as its purpose to inform the practice of teachers. Heyes et al. (1994), attempted to facilitate thinking about the reasons for including drawings as activities in science lessons. One reason is that, in general, drawing activities in the primary school contribute to pupils' sense of enjoyment. The other reasons they place in two categories. One concerns the intention that the activity will lead to the development of individual skills, knowledge, and understanding, that is, "objective purposes." Examples of these are that the children will develop in their ability to observe or gain improved knowledge and understanding of phenomena. The other concerns the teaching and learning processes the teacher has in mind; for example, to facilitate communication between the children or to help keep the teacher informed of the children's thinking. Heyes et al. call these "process purposes." They also suggest that children

> although in the early years of schooling, should be treated as intelligent participants able to comment on their own thinking and responses, with respect to the classroom activity in general, and the drawings in particular. (p. 268)

This means that interviewing the children about their drawing activity should be used as a data gathering strategy, which gives the teacher better information about the children's ideas and understanding than the drawings alone.

Another important issue in the discussion of teaching about the body is the use of metaphors and analogies. Teachers' use of metaphors and analogies while explaining scientific things is discussed in the literature (Ogborn, Kress, Martins, & McGillicuddy, 1996). There it is argued that metaphors and analogies play an important role when teachers are explaining scientific issues. Children's use of metaphorical thinking is discussed by Holgersson (2003). According to Holgersson, children use metaphors more than analogies when trying to explain different scientific phenomena, and to make their explanations more understandable. In order to understand a phenomenon children frequently refer to experiences they find similar to the phenomenon or they construe analogies and metaphors to make things more intelligible (Holgersson, 2003).

2.5   Quiet Children

As seen in the review of the literature on interaction in the classroom (Section 2.3), previous research has mostly been about finding out and taking account of children's ideas and the value of gaining access to children's ideas in order to support and improve their learning. However, according to the experience of Keogh and Naylor (2004), all the guidance about helping children to change and develop their ideas is based on the assumption that children are willing to express their ideas. Keogh and Naylor assume that expression of ideas is important for children's learning, and this view is supported by Harlen (2000) who suggests that learning in science is said to be generally viewed as a process of changing and developing children's ideas but not simply giving them new ideas. Children's ideas therefore have to be taken into account and assessed in the process of helping to improve their learning.

Keogh and Naylor (2004) point out that if children do not feel that their ideas will be valued they are less likely to share them publicly. They say that if teachers want children to "think out loud" they will have to provide the sort of learning environment that makes the children feel comfortable to do so. Children need to know that they are allowed to make mistakes and give wrong answers and that they should respect each other's opinion. Keogh and Naylor's research into the use of *Concept cartoons* indicates that children who are usually reluctant to put forward their ideas or take part in discussion are much more willing to express their ideas when they use concept cartoons. This is especially the case with children with special needs and children who lack confidence. It appears to them that

> using concept cartoons helps provide a learning environment in which children feel comfortable to discuss and argue about what they think, and that is a vital step in helping them to develop their ideas further. (p. 19)

Keogh & Naylor also say that when more traditional methods are used to get access to children's ideas, such as individual true/false statements where each child has to answer on their own on a paper what they think about a specific issue or statement, learners typically feel anxious, insecure, or lacking in confidence. However, when they are working in small groups their responses are generally very different, as they tend to feel more confident, more secure, and less anxious. They also think it is important to get the children to talk in order to develop their ideas and therefore the teacher has to provide an environment for the children to talk and this can be done by using their own concept cartoons (Keogh & Naylor, 2004; Naylor & Keogh, 2000).

Keogh and Naylor developed and published their Concept cartoons in 2000 as a means of representing alternative conceptions in science. Concept cartoons are "cartoon-style drawings which put forward a range of viewpoints about science involved in everyday situation" (Naylor and Keogh, 2000, p. 1). The cartoons are designed to provoke discussion and to stimulate scientific thinking and an approach to teaching, learning, and assessment in science (Naylor & Keogh, 2000). The authors recognize the difficulties of many teachers who try to plan activities on the basis of what the learner already knows and understands but who find it very difficult to manage the constructivist approach in a class of 30+ children (Keogh & Naylor, 1996). With their Concept cartoons they hoped that by presenting alternative ideas in a visually appealing format they would be able to elicit children's ideas and challenge them, leading to their ideas being developed further. Results from their studies show that "the concept cartoons appeared to provide an innovative and effective approach to teaching and learning in science, with considerable potential value for teachers" (Keogh & Naylor, 1996, p. 4). Teachers were very positive and commented on how the concept cartoons promoted cognitive developments in the children, and were particularly useful for encouraging discussion amongst children who were usually reluctant to share their own thinking or get involved in discussion.

The frustration and inability to communicate with and to teach a group of quiet, withdrawn pupils interested Janet Collins in looking into this area (Collins, 1996). The quiet and withdrawn pupils in her class seemed unwilling and unable to join in the social and academic conversation that took place in the class. These children rarely spoke in class and seemed reluctant to ask for help. They adopted a passive role and hardly ever initiated discussions or asked questions. Collins' concern for the quiet pupils stemmed from beliefs that talk is central to children's cognitive and emotional development and that children developed their perception of themselves and the world through talk (Collins, 1996). Her view is supported by Mercer (1995) who talks about what he calls "the guided construction of knowledge" in which one element is "learning" (p. 1 and 2). Mercer describes language as "a social mode of thinking" and builds his ideas on Vygotsky (Mercer, 1995; Vygotsky, 1978).

Teachers assess and support children's learning by talking to them and listening to what they say. For pupils to be successful and make the most of learning opportunities it is important that they become active participants in classroom discussion:

Talk is particularly important in schools, especially at Primary school level when pupils are just beginning to learn to read and write. In these situations talk is the main medium of instruction and assessment. By talking to children and listening to what they have to say teachers assess and support children's learning. For pupils to be successful and make the most of the learning opportunities offered it is important that they become active participants in the discourse of the classroom. (Collins, 1996, p. 2)

Collins' view is also in line with the views expressed earlier (Sections 2.1 and 2.3) that conversations between children and adults greatly aid cognitive development (Burns & Myhill, 2004; Hedegaard, 1999; Myhill, 2003; Ogden, 2000; Vygotsky, 1986).

Collins' attempts to involve the quiet children were not successful, especially in whole-class or large-group discussion. They hardly ever volunteered to answer and if they put up their hand it was just when she had chosen another child to answer. Choosing a quiet child who did not have a hand up seldom succeeded and it caused discomfort and frustration on behalf of the child. In her teaching she presented the children with open-ended tasks and child-centered activities designed to meet the needs of individual pupilsbut she realized that she was not meeting the needs of quiet withdrawn pupils (Collins, 1996).

According to Collins the limitations of her own teacher-directed whole-class discussion could be seen as one of the major factors in her failure to communicate with quiet children, but she notes that group work could also be difficult. When quiet children were put together in a group there was little talk within the group and the feedback session was difficult for them. When quiet children worked in a group with more confident peers it usually led to the quiet pupils being dominated by the confident ones. Therefore quiet children are unlikely to be active participants in the discourse of the classroom, and quiet nonparticipatory behavior can also have a detrimental influence on learning. For confident pupils whole-class discussions give opportunities to demonstrate their knowledge, while quiet pupils are anxious they might be forced to speak against their will. But is it possible to talk about quiet children as a special group? Some children are quiet in the classroombut this may be the only thing they have in common. McCroskey (1980) argues that apart from being quiet they are as different from one another as any other human beings. His view, quite different from that of Collins, is discussed later.

In her study Collins (1996) regarded pupils as being "invisible" when they had no direct contact with their teacher, and quiet withdrawn pupils also tended to occupy a smaller physical space in the classroom than their more dominant peers. According to Collins, however, isolated incidents of "being invisible" do not have to have a significant negative effect on children's learning, as they can be learning even though they are working on their own. "In fact in some situations individuals may benefit from the experience of working independently and without close supervision" (p. 38). Thus, according to Collins there are situations where overt communication does not play a role in learning. Children may learn even though overt communication is not a direct part of a particular activity.

Collins emphasizes that quiet children have to be encouraged to be more assertive and find their voice in the classroom, while the more vocal pupils have to be persuaded to talk less and be an appreciative audience to give the quiet ones a chance to contribute. In the light of her research, she gives examples of teaching strategies for the whole class and emphasizes the need to produce a secure and challenging environment. This is both for social reasons, learning to be a part of social group, and for cognitive reasons. The teaching strategies that she finds most successful are: small-group activities, "talk partners," the "jigsaw strategy" and "going round the circle" activities.

Collins classification of different strategies may be summarized in the following way:

- Small-group activities can be based on friendship groups as chosen by the pupils themselves. Small groups of quiet children can provide children with the opportunity to work with other quiet pupils and develop both social and cognitive skills that they can practice within a relative secure environment before trying them out in a whole-class situation.
- When creating "talk partners," pupils can chose two or three partners that they would like to work with and are allowed to work with one of them. In order to allow the pupils opportunities to work with a larger group, several groups of "talk partners" are put together. "Talk partners" provide pupils with the security they need prior to participating in whole classroom discussion. The "jigsaw strategy" can be used where individuals leave their "home group" in order to gather information from other groups, but this can be distressful for the quiet pupils.
- "Going round the circle" games or strategies proved to be quite successful. A small ball is passed around the circle to indicate whose turn it is to speak. Only the person holding the ball is allowed to speak and the others have to listen. Vocal pupils may find it difficult to remember to remain quiet at appropriate

times. According to Collins, participation in the circle activities is likely to foster a sense of belonging to the class as a community.

- Finally, teacher–pupil "one-to-one interviews" are valuable in providing pupils with an opportunity to talk about their experiences without having to compete with more vocal peers. They can also provide teachers with an opportunity to spend time with and get to know the ideas of pupils who are often overlooked in the classroom. Collins' work emphasizes the importance of developing social skills, in the belief that social skills are necessary for cognitive skills to develop and learning to take place (Collins, 1996).

But could quiet children not be interacting in their own way and participating in their heads even though they are not sharing their thoughts? There is a strong trend in the literature to support the importance of talk and shared discussion in the classroom. According to Harlen "discussion plays a central role in both negotiation of meaning and reflection of thinking" (Harlen, 2006, p. 161). This is supported by the classic work of Barnes (1976), who drew attention to the role of discussion through studying children's speech when they where involved in group tasks. He argued that talking is essential to learning and showed how children contribute to an understanding when talking together about an event, process, or situation. One child's idea is taken up and elaborated by another child, then another adds to it and takes it even further and is perhaps challenged by someone else's idea, which leads them back to check with the evidence. This can only happen if thinking is made open and shared with others through the use of language. When Barnes argues that talk is essential to learning he does not mean formal reporting or direct answering of a teacher's questions, but rather the value of talk among children themselves where they get the opportunity to interrupt each other, hesitate, repeat themselves, and rephrase their words (Barnes, 1976; Harlen, 2006).

Alexander (2004) also discusses the role of talk and describes "dialogic teaching" as "the power of talk to stimulate and extend children's thinking and to advance their learning and understanding" (p. 1). He points out that the process of dialogic teaching is similar to formative assessment in that it enables the teacher to diagnose and assess children's ideas and understanding. It aims to engage children and teachers in active dialogue where listening, responding, asking, answering, expressing and explaining, arguing, justifying, and evaluating is valued and is on-going practice in the classroom (Alexander, 2004; Harlen, 2006).

Collins (1997) notes the importance of talk as a medium of instruction and assessment in primary schools. By talking and listening to what the children have to say the teacher can assess and support pupils' learning. She points out, however, that class or group discussions are often dominated by a small number of confident but not necessarily articulate children:

> A belief that language is central to children's emotional and cognitive development suggests that habitual silence is detrimental to learning. Moreover, non-participation in the social and academic conversation of the classroom prevents children from making the most of the learning opportunities presented to them. (p. 3)

So according to this view, dialogue between pupils and teachers is a vital part of the educational process and pupils that do not have a voice in the classroom are disadvantaged. Silence is a prerequisite for learning, and because quiet behavior does not pose an obvious threat to classroom discipline, the educational and emotional needs of quiet pupils often go undetected. Evidence suggests that quiet withdrawn pupils are often overlooked in busy classrooms.

In yet another study by Collins (1998) she again looks at the issue of quiet pupils but here describes them as "playing truant in mind." The criteria used to select pupils was that they were exhibiting quiet and nonparticipatory behavior in class. Playing truant in mind occurs when a pupil is physically present in the classroom but, for whatever reason, does not participate in the experience that has been planned and presented by the teacher. Ten of twelve pupils fitting these criteria were girls, which raises the issue of possible links between truanting in mind behavior and gender.

This was also the case in a study by Jule (2003) of one classroom where 7-year-old girls displayed consistent silence, both in formal lessons as well as in small groups in over forty hours of classroom observation. There were twenty pupils in the class, eleven boys, and nine girls. Results showed that the boys used significantly more linguistic space than the girls, expressing themselves more and talking more than the girls. According to Gurian (2001), boys tend to be louder and more prone to employing attention-getting devices in the classrooms than girls, which results in more attention given to boys. In some classrooms boys dominate discussion and girls' voices are lost. We should, however, be wary of portraying "boys" and "girls" as homogeneous groups. A generation ago the under-achievement of girls was identified in schools (Bleach, 1998); but now the challenge is reversed in a number of countries because girls' achievements in education have risen markedly, and the concern is now for the "lost boys." According to Bleach, girls tend to have "a compliant motivational style" while boys want to do everything more quickly and prefer short tasks; boys take less care with the standard of their work, are reluctant to do extra work, read less and are also less attentive in class, and their manner is often lively, though this of course is not true for all boys.

In a research by Tunnicliffe and Reiss (1999c) on how children see animals they discuss gender difference: According to their results, girls used a richer variety of explanations than boys who largely relied on anatomical consideration alone. The girls appeared to be more likely to comment on features of the animals' faces, suggesting that they are more likely to show empathy than boys. But the boys were more likely than the girls to cite books as sources of knowledge (p. 147).

It is difficult to know what is going on in the child's mind and to know the extent to which the pupil understands certain text or information. This can only be ascertained by talking to the child and/or by assessing their ability to complete a task which requires knowledge from the text or information given. Collins (1998) identified four types of withdrawal or truanting behavior exhibited by quiet pupils: "being invisible," "refusing to participate," "hesitation," and "inappropriate focus." "Being invisible" means that pupils avoid having direct contact with the teacher during lesson. Where they sit and how they behave can make them "invisible" and minimize their contact with the teacher. Alternatively, pupils would be invited to participate in discussion or an activity but would "refuse" to join in. Sometimes there could be a valid reason for a pupil not wanting to join in, but in other cases they would just remain still and quiet and avoid making eye contact with the teacher. According to Collins, these two types are easy to detect but the other two are more difficult because in both situations the pupil can appear to be busy but closer analysis shows that they are not actively engaged in the tasks set by the teacher. In the third type of truancy, "hesitation," the pupils appeared busy but they never became engaged in the task and sometimes seemed afraid of participating. The fourth and the final type was when pupils had "an appropriate focus" and according to Collins this was the most disturbing one as the pupils here would take part in an activity that had little or no connection with the learning tasks presented by the teacher.

There are a number of issues that seem to have a significant effect on truanting in mind behavior: teacher expectation, feelings of security, teaching style, and classroom organisation. These issues seem more significant than teacher popularity or the subject being taught. Collins (1996) suggests that, to reduce truancy in mind, the strategy of "maintaining and monitoring attendance" would have to be interpreted in terms of organizing and managing the classroom in such a way as to maximize participation, to provide direct support for the emotional and behavioral needs of those who play truant in mind, and to offer alternative curriculum experiences.

Other studies that discuss the dilemma of involving all students in whole classroom discussion mention similar strategies as Collins does in order to get pupils involved (Hu & Fell-Eisenkraft, 2003; Osborne, 1997). In a study on immigrant Chinese students in an American school, Hu and Fell-Eisenkraft (2003) suggest similar teaching strategies as Collins (1996) for involving children in classroom discussion, for example, the "jigsaw strategy" and something that they call "fishbowl discussion," which can be used when one or two students dominate the discussion. In "fishbowl discussion" children in a group that is already discussing a particular issue are selected to continue their conversation while the rest of the class gathers in a circle around them. Then each student in the outside circle is assigned to observe one student in the discussion group. The observers look for both active and passive interaction. The fishbowls "provide a good opportunity for students to self-evaluate and articulate individual and group goals for working towards more effective, rewarding discussion" (Hu & Fell-Eisenkraft, 2003, p. 59).

Osborne (1997) stresses that not only the individuality of the child but also the individuals' abilities to work within a group are extremely important, maintaining that the role of the teacher is to create conditions for learning where children can act as individuals but also as a group. Osborne argues that there is an inherent dilemma in constructivist teaching between serving the needs of the individual child and that of the class. During the course of her 3-year study in her own class of first, second, and third grade, Osborne analyzed the role of the individual, the group, and the teacher in a constructivist classroom and discusses how the contribution of a particular child can affect the entire class. One boy in the class was very active and energetic, funny, and creative and used to dominate the discussion. He was a leader and the other children and the teacher loved to be around him. He was very observant and knowledgeable about the way things worked and was curious and enthusiastic. He was happy to talk and he made very helpful and stimulating comments that the other children could hear and were used to challenging ideas and moving the discussion onto new things. He was an extremely important part of the teacher's science instruction. The insights he contributed to the class in science were invaluable and so was his participation in various tasks. But he dominated the discussion and it is difficult to know what the others thought. And the question is again, were the quiet children participating in their heads or where they some place else? In this sort of teaching environment the teacher is dependent on children thinking out loud although participation cannot be forced on the child (Osborne, 1997).

A study by Lomangino et al. (1999) emphasizes the importance of encouraging children to share their ideas and acknowledge the ideas of others.

Getting ideas from all participants within a group of children may diminish the tendency for high-status individuals to dominate, while low-status children are blocked from participation. They support both Piaget's and Vygotsky's theories of development when referring to the potential benefits of collaborative activity and say that within their contrasting perspectives on development, both Piaget and Vygotsky present the importance of interaction with others for learning.

Quiet children have only one thing in common, they are quiet. Beyond that they are as different from one another as any other human beings (McCroskey, 1980). McCroskey argues that most quiet children do not have a problem and teachers should not try to change the young person's personality, but instead try to take positive steps to involve the quiet person in discussion. If communication with other children in the class is unrestricted it is much more likely that a pupil, even a quiet one, will engage in communication. Such a climate is developed when a teacher reinforces pupils in communicating with others. While this type of classroom atmosphere also encourages some conversations which are not directly conductive to learning, the overall impact is supportive of the learning process for all pupils, not just the quiet ones (McCroskey, 1980). This view was also argued for by Barnes (1976) when he said that talking is essential to learning and that nonformal discussion amongst the children is the most effective sort.

The use of discussion in the classroom is a valid and important instructional strategy for the teacher and for the learner, as "discussion plays a central role in both negotiation of meaning and reflection on thinking" and it is through language that we develop a shared understanding of ideas (Harlen, 2006, p. 161). However, what is valuable and beneficial to some children is not necessarily so to others. Forcing pupils to perform orally can be harmful as it can increase apprehension and reduce self-esteem. The quiet pupil is put in a weak position when oral performance is the only way to demonstrate achievement. Therefore, McCroskey (1980) argues, the teacher should permit and encourage oral performance and discussion but never require it from quiet pupils. It is hardly ever necessary for a pupil to perform orally to demonstrate learning, and the teacher should develop alternative methods for the pupil to demonstrate achievement, allowing pupils the choice of oral or written forms of demonstration and evaluation should be based on what the pupil knows, not on how much he talks.

Boshell (1995) discusses what can be learned by giving quiet children space. In his research there were a number of quiet children in the class whose participation was rather limited. These children tended either to give very short responses or to start off explaining something but then stop and not complete the explanation. Each of the quiet children was asked why their contribution was so limited: Was it because they had not understood what was being discussed? Most of the children commented that they had understood, although their participation tended to be limited in the class.

According to Holbrook (1987) some teachers may regard the quiet child as "perfect" in that they are not discipline problems. But often the quiet pupils' lack of response or participation has a negative effect on how they are seen by the teacher and they are perceived as less capable and are thus called on less frequently in class discussion. Their lack of enthusiasm tends to limit teachers' attention to them, which further reinforces their own perhaps negative self-evaluation. The school environment can play a vital role in preventing communication apprehension. In Holbrook's view, the characteristics of a healthy classroom include: creating a warm, easy-going climate in the classroom; helping the children get to know one another at the beginning of the year; using drama and role-play situations; having students speak to the class in groups or panels rather than individually; allowing students to work with classmates with whom they feel most comfortable; and having students speak from their seats rather than from the front of the classroom (Holbrook, 1987).

Harlen and Qualter (2004) discuss gaining access to children's ideas. "Children have to feel free to express their own ideas and ways of thinking, without fear that they will be giving the "wrong" answer" (p. 141). According to them the teacher has therefore to establish a classroom climate in which children feel safe to express their ideas, and various strategies can be used to get access to children's ideas such as questioning, writings, drawings, concept maps, concept cartoons, and discussion.

Harlen and Qualter also say it is usually implicit that questioning and answering strategies are oral. In some circumstances it may be better for children to write down their ideas or to produce a drawing or set of drawings to express their ideas. This can give the teacher information about the full range of ideas in the class and a permanent record of the ideas of each child, which can be looked at and used later. Concept maps can be useful in finding out children's ideas where conceptual links are put between words (Harlen & Qualter, 2004). If we use the words "ice" and "water" we can connect them with an arrow to signify a link between them and then write a word that indicates what happens between the two concepts, like ice "melts to give" water.

Thus, there is a strong trend in the literature to support the importance of talk and shared discussion in the classroom and talk is seen as an important contribution to understanding. The fact that some children are reluctant to talk in the classroom and share their ideas is seen by most authors as a problem but not all. A number of ways have been put forward to get access to such children's ideas and suggestions have also been put forward that aim at encouraging quiet children to share their ideas.

2.6 Research Questions

In this review of previous research an attempt has been made to look at different aspects of the constructivist view of learning. Special attention has been given to Piaget's and Vygotsky's contributions and to the differences between their ideas but also to the similarities and to what can be learned and used from their important contributions to education. The power of egocentric speech has been discussed and also the idea of scaffolding as a means of supporting children's learning. Constructivism and teaching have also been reviewed and it has been discussed what constructivism has to offer in respect to the role of the teacher and the importance of certain types of teaching environment. Although there is no specific prescription of constructivist teaching, some ideas are put forward in the literature that are thought to fit well with the constructivist view, but this is an area that needs to be explored further. The term "active" in referring to learning in the classroom has also been examined, but also needs to be explored further.

From the literature review there is an emerging picture, both of the understanding that young children have about the human body, and under what external conditions it develops. Special emphasis has been put on interaction in the classroom and the dilemma of how to elicit the ideas of all the children in a class, including those of quiet children, and how to engage all the children in class discussion. The studies that have been referred to above do not, however, answer all the questions about the influence that teaching and interaction have on this development: There are disagreements between some of them and it is therefore difficult to conclude from them whether overt verbal activity is a prerequisite for conceptual learning in biology or whether it is very important for some children but not for others. From the literature, in particular, we do not know in much detail how the ideas and understanding of the location, structures, and the functions of the different organs of the human body develop in young school children.

As a result of my overall research interests (Chapter 1) and the above literature review (Chapter 2), especially the literature about the change in children's ideas and the influence the environment has on the change and development of ideas, it was decided that the following research questions would be formulated.

1. How do the ideas that Icelandic children bring to primary school change over the course of the first two school years during teaching about the human body in relation to location and structure (bones, muscles, heart, lungs, brain, digestive-system), function (of the heart, skeleton, lungs, brain, stomach), and process (digestion and blood circulation)? This question invited the subquestion: What kinds of ideas do children in the 1st grade of primary school in Iceland (six years old) have about their body before teaching about the human body begins?

Previous research does not tell us enough about how teaching methods, teaching materials, and different tasks affect the development of the children's ideas, nor do we know enough about the influence of teacher–pupil interaction and peer interaction in the development of children's ideas about the human body. Furthermore, we know very little about the difference between quiet children's learning and that of the more open children, or even if there is much of a difference. To obtain information on these important issues the following additional research questions were formulated:

2. How are changes in pupils' ideas affected by: the curriculum, teaching methods, teaching materials, teacher–pupil and peer interactions, or other factors?
3. What are the differences between quiet children and more open children in respect to these issues?

These research questions were refined throughout the course of the study. Indeed, the focus on quiet children only emerged as a result of the initial stages of fieldwork.

3

Methodology

This chapter describes the setting in which the research took place, the participants, and the methods that were used to collect and analyze data. The research took place in a primary school in Reykjavik from the 14th of February 2003 until the 28th of February 2004. There was one class of children involved in the study along with its teacher and a sample of parents. The children were 20 in number in Primary 1 (Age 6) when the research started but 19 in Primary 2 (Age 7) when data collection finished. The Event Diagram (Table 3.1) shows when different content was taught and I observed all these instances and when different additional data was collected.

According to the *National Curriculum Guide: Natural Science* (Menntamálaráðuneytið, 1999) the curriculum should be organized in a spiral manner so that the students continually build upon what they have already learned, which is in tune with Bruner's (1996) notion of a "spiral curriculum," where basic ideas should be revisited and built upon. The curricular episodes are organized with this in mind. Therefore Episode 7–9 issues are explored in more detail and depth than in Episode 1–6.

"The Brain Controls Everything", pages 63–96
Copyright © 2016 by Information Age Publishing
All rights of reproduction in any form reserved.

TABLE 3.1   Event Diagram for the Study

Primary 1: 14th of February–End of May 2003 (20 children)

Curricular Episode 1: What Do We Know About Our Bodies? Bones and Muscles
1. Exploratory phase: Meeting with the teachers.
2. The teacher explores what the children know about the body. Classroom observation.
3. Initial drawings of the bones and the organs. Drawings 1 and 2.
4. Meeting with the teachers.
5. Teaching about the bones, muscles, and joints. Discussion about the external body parts and the organs the children know. Classroom observation.
6. Second drawings of the bones and the organs (two weeks later). Drawings 3 and 4.
7. Interview with the teacher.

Curricular Episode 2: The Heart is a Muscle That Pumps Blood
1. Teaching about the heart (not blood circulation). Classroom observation.
2. Coloring a detailed drawing of the heart from a three-dimensional model. Drawing 5.

Curricular Episode 3: Healthy Living, Healthy Food, and the Food in the Stomach; Muscles
1. Meeting with the teachers.
2. The initial drawing of the food in the stomach. Drawing 6.
3. Muscles that we can control and the ones that we cannot control (e.g., in the stomach and the heart). The SHIPS activity about the "Moving feet" sent home with the children. Classroom observation.
4. Drawing of the muscles on a picture of the skeleton. Drawing 7.
5. Meeting with the teachers.

Primary 2: September 2003–February 2004 (19 children)

Curricular Episode 4: Reproduction
1. Meeting with the teachers.
2. Teaching about reproduction: discussion, pictures from Let's look at the body. Classroom observation.

Curricular Episode 5. The Skin, Senses and the Nerve-Cells
1. Meeting with the teachers.
2. Teaching about the skin, senses and the nerve-cells. Classroom observation.

Curricular Episode 6: The Heart, Lungs, and Blood Circulation
1. Teaching about the heart, lungs and blood circulation. Classroom observation.
2. The children did a drawing of the heart, lungs and blood circulation. Drawing 8.
3. Interview with the teacher.

Curricular Episode 7: The Stomach, the Digestion System, Liver, and Kidneys
1. Teaching about the stomach, the digestion system, liver, and kidneys. Classroom observation.
2. Second drawing of the food in the stomach. Drawing 9.

Curricular Episode 8: The brain
1. Teaching about the brain. Classroom observation.
2. The children drew and colored the brain. Drawing 10.
3. Interview with the teacher.

(continued)

TABLE 3.1   Event Diagram for the Study (continued)

Final Phase
  1. Individual interviews with the children.
  2. The children completed diagnostic tasks.
  3. Interviews with the parents.

When thinking of a writing style to describe how I collected and analyzed the data, I decided, after giving it a lot of thought, to use a personal style where I use the first person. Here I am inspired by the style Michael Reiss uses in his book *Understanding Science Lessons* (2000). His description of how he carried out his longitudinal study into pupils' learning of science is very personal and I like the style he uses. I think by using this style he helps the reader get both a deeper and also a broader perspective of how the research was carried out. However, I am aware that being personal is not the traditional research style of reporting and that those readers who prefer a more conventional style can find the personal style overwhelming. With this in mind, I try to keep the personal style within limits, but there are sections where I find it very important to use the first person in order to give the reader both a broader and a deeper insight into how and why I carried out the research the way I did.

3.1   Research Design

The research design used has some elements of "one group pre-test/post-test design" where the concern is to determine whether there is an increase or change of performance after a treatment, a single group is pretested, gets treatment, and is tested again (Cohen, Manion, & Morrison, 2000; Robson, 2002). This design can be used in a qualitative case study with multiple sources of evidence where, in addition, some quantitative pre- and post-intervention data are collected on a small number of variables (Robson, 2002). The research focus in this study is on one class/group of children and their ideas about the body before working through an educational project about the human body and on how their ideas developed over a period of one year. The children made drawings before the teaching started (pre-test) and also during the year's teaching about the body. At the end, they were interviewed and also asked to complete some diagnostic tests (post-test). This design is, however, open to the accusation that it is vulnerable to many threats to validity. Other events could have had an influence on the children's ideas apart from the teaching in the classroom between the pre- and post-tests and maturation could also have had an affect through developments in the group during this period (Cohen et al., 2000; Robson, 2002). This research is not a piece of "experimental research" involving an "experimental treatment" between pre- and post-tests, but it has elements from it as described here which may be useful to the interpretive and subjective research approach used.

3.2 The Tradition Within Which the Research Is Situated and the Methods Used

Overall, the research tradition can be described as "eclectic" since a number of research methods were used, as discussed below, in order to provide information from different angles and so increase the validity of the conclusions. The methods used were classroom observations; observations of teachers' meetings, interviews with the teacher, the children and a sample of parents', and children's drawings; and the results of a range of diagnostic tasks undertaken by the children. This chapter explains how and why these different methods were used and discusses the advantages and disadvantages of each.

This research can be placed within two of what some describe as the five main traditions of qualitative inquiry (Bogdan & Biklen, 2003; Creswell, 1998; Denzin & Lincon, 1998). For a start, the research has elements of ethnography. Ethnography focuses on a group of people that has something in common, usually a culture-sharing group, and has its original roots in anthropology. Data are collected by observation (participant or nonparticipant) and through open interviews (Creswell, 1998; Denzin & Lincon, 1998; Emerson, Fretz, & Shaw, 1995; Wolcott, 1994). Ethnography is an open approach and more inductive than deductive, and the term is usually understood in two ways, as a method of investigation and as a text (Marinosson, 1998; Van Maanen, 1996). As a method, it refers to participant observation where a single researcher, over a particular length of time, is studying a particular culture. As a text it refers to writing a representation of that culture: "a detailed enough description of an event or a situation to convey the meaning it carries in its context" (Marinosson, 1998, p. 35).

Traditional ethnographers seek to explain cultures by understanding the members of that culture. The ethnographer becomes a member of the group and reports their life on their terms. However, the subject's behavior is always explained in terms of the ethnographer's point of view and understanding (Fontana, 1994): "Thus, despite the researcher's claim to be "invisible" in the report, he is dominant, for all that is reported is filtered through his eyes, his heart, and his mind" (p. 211).

More recently postmodern ethnographers have become much more "visible" and admit that they have influenced the study in some degree, that the ethnographer becomes a visible partner in dialogue and in the data and this acknowledgement of visibility is aimed at reducing the ethnographer's authorial influence on the data (Fontana, 1994). Thus, postmodernist reflexivity has dramatically altered traditional ethnography: There is no longer a dominant mode of ethnography but instead a wide range of new postmodern approaches although traditional ethnography continues as the prominent mode of field research. Postmodern fieldwork emphasizes the status of the researcher as the subjective author of ethnographic accounts and it relies on awareness of problems in the field, but still bases its observations on everyday data gathered from the people being studied, the "natives." The concept of "everyday life" is also seen more broadly and can take into account, for example, films, television, dreams, fiction, and other types of data that have not commonly been included by traditional ethnographers as part of their field inquiry. This postmodern trend has therefore been called "multitextual ethnography" because it uses a number of different sources and:

> in questioning their own procedures by rendering problematic the authorial subject and in broadening the objective field of enquiry. Seen in this way, what makes ethnographies postmodern is their questioning of traditional ethnographic modes that have been mired in paradigmatic stagnation. Postmodern ethnographers do not advocate any one new way of doing and reporting ethnography; instead, they favour a multiplicity of approaches. (Fontana, 1994, p. 220)

This research is a type of multi-textual ethnography: It focuses on a group within that group's own context of social interaction (being in the same class at school) and studies the meanings of behavior, language, and interaction of the group within the setting. Data are collected through observation, partly participant observation, and interviewing. Teachers' meetings, for example, were observed where the focus was on how decisions were made, on how the teachers discussed and decided on issues and activities, and on how they planned lessons. There was also extensive classroom observation where the behavior, language, and interactions of the "culture-sharing group," namely the class of children, were studied, while the teacher shared in collecting other types of data such as drawings and written tasks.

Some of the important elements of the case study approach can also be identified in this research, that is, the research was substantially bounded by time and place (Creswell, 1998). It has its time, one year, where a special educational project is observed and evaluated and the people studied (here children) are in one group and in a particular context and setting, a particular class in a particular classroom. Extensive, multiple sources of information are used in data collection to provide a detailed, in-depth picture of the classroom interaction, the curriculum, and the teaching and learning that takes place (Bogdan & Biklen, 2003; Creswell, 1998; Robson, 2002; Yin, 2003). Case studies are a special kind of qualitative work, which investigates one or more contextualized contemporary "instances" that have specific boundaries. Examples of such bounded "cases" or phenomena within education can be a program, a setting, one child, or a social group (Hatch, 2002; Merriam, 1988; Yin, 2003).

The term "case study" draws attention to the question of what can be learned from a single case. A case study optimizes understanding of the case rather than generalization beyond that particular study. "The purpose of case study is not to represent the world, but to represent the case" (Stake, 1998, p. 104). A case can be simple or complex: one child or a classroom of children, or a study of classroom conditions. The time spent on our study can also be short or long depending on each individual case. The case is also a specific and integrated system that has its own boundaries, "but the boundedness and the behaviour patterns of the system are key factors in understanding the case" (p. 87) and each case study is a concentrated inquiry into a single case. Stake identifies three types of case study. This study can be identified with the type he calls "intrinsic case study," which is a study undertaken because the researcher is interested in better understanding of a particular case. It is not undertaken because it represents other cases, but because of its particularity and ordinariness the case is of interest itself and may reveal its own story. Furthermore, such revelations can shine light on other cases.

This study fits well into this description and the ideas discussed here. Like postmodern ethnographic studies, case studies are also multi-textual, and ethnographic methods fit well with the case study approach where different methods and sources are used to obtain information. Both the case study and the post-modern ethnographic approach have the ethos of interpretive study and aim at seeking out the common meanings held by the people within the case. The case study researcher like the ethnographer spends a substantial amount of time on site, personally in contact with the operations and activities of the case, reflecting and revising meanings of what is going on (Stake, 1998). The ethnographic case study researcher seeks to see what is natural in happenings and in settings, but what he is unable to see himself is obtained by interviewing people or getting information by other means such as documents and written material (in this study, drawings and written tasks). It is not clear from the research literature to what extent qualitative case study research is viewed as being distinct from ethnography or participant observation study (Hatch, 2002). Indeed, some authors do not use such widely accepted terms as: qualitative, ethnographic, case study, participant observation, and phenomenological, but prefer to use the term "interpretive" approaches to cover all of them. "To think of research as interpretive also reminds one that all research is about interpreting data records and making those interpretations public" (Graue & Walsh, 1998, p. 17).

To reduce the likelihood of misinterpretation it is valuable to employ various procedures called triangulation. "Triangulation has been generally considered a process of using multiple perceptions to clarify meaning, verifying the repeatability of an observation or interpretation" (Stake, 1998, p. 97). Denzin and Lincon (1998) identified the following four basic types of triangulation:

1. Data Triangulation (the use of a variety of data sources in a study)
2. Investigator Triangulation (the use of several different researchers or evaluators)
3. Theory Triangulation (the use of multiple perspectives to interpret a single set of data)
4. Methodological Triangulation (the use of multiple methods to study a single problem)

The fourth and the first of Denzin and Lincon's types describe the approach that was adopted in this study as different methods were used to collect a variety of data. Data produced by different methods can be compared: For example, observational data can be compared with interview data (Foster, 1996). Delamont (2002) talks about triangulation "between" methods and "within" a method, where "between methods" means gathering data by more than one method and "within method" triangulation means systematically attempting to obtain several types of data within the method you are using. Triangulation has also been identified as "crystallization," a term used by Richardson (1994). He explains the idea of crystallization as a lens that can be used to view research designs and their components. According to him the "crystal combines symmetry and substance with an infinite variety of shapes, substances, transmutations, multi-dimensionalities, and angles of approach. Crystals grow, change and alter, but are not amorphous" (p. 522). Janesick (2003) favors this view and says that what we see when we view a crystal depends on how we view it and whether we hold it up to the light or not. According to her, crystallization incorporates the use of other disciplines to inform our research processes and to broaden our understanding of method and substance. So whatever word is used, triangulation or crystallization is a way of checking and cross checking data, findings, methods, and other aspects of the research to be more confident about its validity.

I myself am also bound to have an influence on the situation and then on the research itself since I am the one who collects, analyses, and interprets the data. My experience, my views, and beliefs are bound to have at least some effect on the interpretation although I am aware of the importance of trying not to influence the situation too much and being as transparent as possible but I am still the one who discusses and interprets the finding. This is where triangulation comes in and also "reflexivity," which involves the researcher monitoring and reflecting on his or her own influence, and on the effect of social context, on data production. The researcher is constantly scrutinizing the production of data, considering potential sources of error and evaluating his or her own role. Where such an approach has been consciously adopted we can be more confident that data and research findings are free from "error" as much as possible (Foster, 1996). In Michael Reiss's (2000) longitudinal study into pupils' learning of science, he discusses how he tried to reflect on how he felt about certain things and issues during the course of the study and how this influenced his interpretations. He says "What I write is inevitably shaped by whom I am" (p. 13). If someone else had done the same study it would have been different: "A study such as this cannot be fully objective. For various reasons, the observations made and conclusions reached by another researcher would have differed from mine in at least some respect" (p. 15).

In this study, different methods were used to analyze the data. Data analysis is a systematic search for meaning and a way to process data so that the findings from the data can be communicated to others. Analysis also means organizing and interrogating data so as to allow the researcher to see patterns, identify themes, discover relationships, and develop explanations (Hatch, 2002; Lofland & Lofland, 1995). This can take three main forms: description, evaluation, and explanation. Description involves combining data in order to identify the key features of the phenomena of interest. Evaluation involves comparing descriptive data, and explanations involves looking for connections and relationships among data to establish the cause of particular features or patterns (Foster, 1996). "The main analytic tasks are related to establishing patterns or regularities in the data, and then cross-checking to make sure the data are reliable and valid" (Delamont, 2002, p. 180).

In qualitative field studies, analysis is conceived as an emergent product and the researcher's way of ordering of the data. Even though there are several concrete and even routine activities involved in analyzing data, the process is intended to be open-ended in character. So in that sense analysis is also very much a creative act and there is no single way to analyze (Lofland & Lofland, 1995). However, the main methods used here are extensively discussed in the research literature.

"Discourse analysis" is used to give a clearer picture of the use of certain words or phrases as used by the teacher. Discourse analysis is concerned with the ways in which meaning is reproduced and transformed (Banister & Parker, 1994; Kvale, 1996; Robson, 2002). Conversation can be considered as an access to knowledge (Kvale, 1996). Kvale talks about three forms of conversation: everyday interactions, professional interchange, and philosophical dialogue. These three, he says, may be seen as an oral exchange of opinions, ideas, and observations and involve different forms of interaction. Everyday interaction may be on a simple topic of conversation. Professional interchange can be job interviews, therapeutic interviews, or research interviews. But each of these has a different purpose and structure. In professional interviews there is usually asymmetry of power in that the professional is in charge of the questioning. In philosophical dialogue the partners are on more equal level and the discourse rests on a joint commitment of the participants to seek the truth. According to Kvale, the ideal style of conversation pertains to the philosophical discourse but may also in some cases apply to everyday interactions.

Burman and Parker (1993) discuss the advantages of discourse analysis and outline the key reference points for the development of the method. Discourse analysis is used when studying spoken or written texts and conversations; it is not a single approach but a group of different approaches that are "united by common attention to the significance and structuring effects of language and are associated with interpretive and reflexive styles of analysis" (p. 234).

I used elements from grounded theory to find categories and themes in my data relating to certain issues. Grounded theorists give priority to developing rather than to verifying analytic propositions. By making frequent comparisons across the data, the researcher can modify, develop, and extend theoretical propositions so that they fit the data. At the working level, the researcher begins coding data in close, systematic ways so he can generate analytic categories (Creswell, 1998; Emerson et al., 1995). Constant comparison engages the researcher in inductive and deductive thinking and potential categories of meaning emerge from the data; these categories are continuously and carefully examined to determine if they are valid (Hatch, 2002).

I also used hermeneutic methods to analyze the data; hermeneutic methods are situated in a constructivist approach (Hatch, 2002). This implies that the meaning individuals give to their experiences ought to be the objects of study (Bogdan & Biklen, 2003). According to Kvale (1996) the purpose of hermeneutical interpretation is to obtain a valid and common understanding of the meaning of a text. The subject matter of classical hermeneutics was the text of literature or religion but this has now been extended and "text" can be discourse or even action. The research interview is a conversation with oral discourse transformed into texts to be interpreted (Kvale, 1996). I looked through all my notes and transcripts and also through all the drawings the children made throughout the project seeking signs of certain issues through the use of different teaching methods, teaching material, and incidents. I looked for all the traces relevant to certain issues and I tried to formulate some interpretations on the basis of the material and then tried to see whether the data supported the interpretation (Gustavson, 1996).

When analyzing goes on alongside data collection, as it did in this research, coding and writing memos "are the core physical activities of developing analysis" (Lofland & Lofland, 1995 p. 186). The word used to label or classify data or items of information is a "code" and coding, which can be used in almost every type of analysis, and begins with the process of sorting and categorizing data. "Memos" are the written explanations or expression of what a particular code is about and the relationship between codes (Lofland & Lofland, 1995; Miles & Hubermann, 1994). All these methods overlap and I find it useful to use all of them in order to obtain a better understanding and get a more holistic picture of the issues and phenomena that I am exploring.

3.3 The Setting

One primary school in Reykjavík was chosen to take part in the research. I had worked with the teachers in the school on another project—the SHIPS project (*School Home Investigations in Primary Science* (Solomon & Lee, 1991) and knew that the teachers of Primary 1 were starting teaching about the human body using the teaching material *Komdu og Skoðaðu Líkamann/Let's Look at the Body* (Óskarsdóttir & Hermannsdóttir, 2001a). I thought it helpful that the school was known for good science teaching: The teachers had taken part in courses and development projects in primary science and had been nominated by the "Home and School Association" for a prize for using the SHIPS project (Solomon & Lee, 1991) as a way of establishing a link between home and school. The school was also one of Reykjavik's "mother" schools in science, which means that it is a model for other schools. The teachers in the school had shown that they were open to trying new things and were likely to be motivated and willing to take part in research like this, as turned out to be the case. The head teacher was very keen when I asked if I could do the research in the school. She thought it a privilege and was willing to assist me in every way. So right from the beginning I felt myself very welcome and at home in the school.

Four teachers were teaching the four Primary 1 classes in the school. At the beginning I thought it would be important to have all of them involved in some way, but I subsequently decided to focus on just one teacher and her class especially. The reason why I chose that particular teacher was that I had known her personally for a long time, and I knew she was very interested in primary science and an enthusiastic science teacher. In addition, she was comfortable with having me watch her teach and from the start she was keen to let me do the research in her class. She was an official leader in science teaching in the teacher group. In her Primary 1 class there were 20 pupils—10 boys and 10 girls. The research continued into Primary 2 but by then three children had moved away and another two had joined, so by the end of the fieldwork there were nineteen in all (10 boys and 9 girls).

All four teachers that taught Primary 1 had wide professional experience. They had all worked in that particular school for the whole of their teaching lives, from 8–20 years, and three, including the teacher I focused on, had attended in-service courses on the teaching of primary science.

Parents of six children were interviewed in order to see if there was something especial in relation to the child's interest in the body. I decided that six parents were enough to obtain some information about this issue. I chose the parents of two children, a boy and a girl, the two (of each gender) that contributed most to class discussion, but the teacher picked the rest, that is the parents of two girls and two boys by random.

I started my fieldwork on the 14th of February 2003 after I had completed the formalities and obtained ethical clearance. I sent a letter to Persónuvernd (The Icelandic Data Protection Authority) and to the Educational Office in Reykjavik for permission for the research. I also sent a formal letter to the head teacher of the school and to the parents of the children in the class informing them about the research and asking them for permission to do the research in the school/class. I also sent letters to the parents asking if their children could be interviewed, with the request to sign an agreement slip and send it back with their child to school. No one refused to allow their child to take part or be interviewed.

3.4 The Teacher

The teacher in focus has a BEd degree and also a BA degree in linguistics. She lived in the United States with her family for a few years in the 1990s and took part in her own children's school activities, both by helping them with their homework and also by coming to the school and helping out there. She became very interested in primary science education during her stay in the United States, bought books and journals and collected many ideas that she then brought with her to Iceland and has subsequently used in her own teaching. She said that she had seen in the States how interesting science education could be both for the children and also for the teacher and it really got her interested in science and science education. Since her return from the United States, she has taught in the school where the research took place. She has attended lectures and inservice courses on primary science and has been the science coordinator for the early primary classes in her school.

3.5 The Classroom

The classroom is a traditional classroom in one of the oldest schools in Reykjavík, established in 1946. It is bright, with big windows on one of its four sides. The windows face the school's playing area. In the front is the blackboard and the teacher's table (used mainly as a storage table: I never saw the teacher sit at the table). There are shelves and storage cupboards for material and paper on both sides of the blackboard and also at the back of the classroom. At the back there are also bookshelves for books, both educational material books and books from the school library. There is also a cupboard with containers and boxes for each child marked with their names for their individual work.

The walls were covered with drawings, paintings, and work made by the children, for example, paper dolls of themselves, birthday calendars made by the children, and also some posters and pictures that the teacher had hung on the wall, like a weather teddy and the alphabet. There was hardly an empty space on the walls. There were two computers in the classroom, one at the back and one at the front, both with Internet access. The children were rotated and moved frequently so they got to know each other, changing seats every two weeks. Usually they sat in pairs but sometimes in groups of four and sometimes all the tables were arranged in U shape. Sometimes girls and boys sat separately and sometimes together. Arrangements also changed according to tasks and sometimes the corridor, the stairs, and the science classroom were used. The science classroom is equipped with a lot of science material and tools for use in science lessons for the whole school and the teachers have to prebook it before using it.

3.6   Classroom Observation

Hatch (2002) uses the term "observation" to describe "a specific data collection strategy that can be applied across many kinds of qualitative studies" (p. 72) and the kind of observation that is most commonly used in qualitative studies is usually called "participant observation" because the researcher usually acts at some level as a participant in the setting he or she is studying. The researcher's direct observation permits better understanding of the context they are studying and the experience allows them to be open to discovering inductively how the people in the study understand or make sense of the setting or the situation. And not least, the researcher can learn sensitive information from observation that cannot easily be gained through interview (Hatch, 2002). However, observation and interviews are approaches often used together so as to complement each other, and informal interviews and conversations are often interwoven with observation (Merriam, 1988) as was the case in this study. The level of participation that a researcher takes in the research setting is an important issue. Classic ethnographers assumed that observation meant sharing and participating in the daily life of the people they were studying. But observation as a research method can range between extremes from "complete observer" to "complete participant" or "limited observer" to "active participant" and everything in between. The more involved the observer is in the setting, the closer he or she usually is in the action (Hatch, 2002; Merriam, 1988). An observer who has decided not to participate at all is able to take notes in an educational setting without being distracted by children asking for help or being more than occasionally curious about what he or she is doing, and may, eventually, not have to worry unduly about the effect he or she has on the setting, the group, or the child. Full participation makes it problematic to take field notes, and one's memory is limited, so writing good notes is a challenge. However, the participant observer is able to hear and see, interact, and share to some extent (Graue & Walsh, 1998).

In undertaking research that has strong elements of case study and/or ethnographic approaches, participant observation is usually an important part of gathering the information needed, and the researcher may actually participate in the events being studied (Yin, 2003). But still, researchers have to decide about their level of involvement and have to take into consideration how their participation could influence the natural flow of the events within the setting and context. People involved in the study should know that the researcher is a researcher even though he or she sometimes takes part in the discussion or acts like a teacher (Hatch, 2002). Care needs to be taken in presenting the research topic and the researcher to the children. Researchers doing classroom research need to introduce themselves as someone who wants to "find out about more" about certain things (Mauthner, 1997).

I observed 26 lessons, all the lessons in Primary 1 and 2 when the teacher was teaching about the human body. Each lesson is 40 minutes, sometimes a little less, sometimes a little more, because the timetable in Icelandic Primary schools is very flexible. I videotaped all the lessons and I sometimes also took field notes. I decided to videotape the lessons to have a permanent and detailed record of who said what and when, and in what circumstances. In addition, and because the research is about children's ideas and how they change, I wanted to be able to go back to the videos and look for issues or words that perhaps might have an effect or influence on the children's ideas. I also did not have much time for writing up the field notes immediately after the lessons, so in the absence of videotaping I would not have had an accurate record of all the important words and incidents that possibly had some influence on the development of ideas. Graue and Walsh (1998) and Hatch (2002) discuss the strengths and weaknesses of video recording as a research tool. According to Hatch (2002) "video recording can provide a powerful means for capturing data that can improve the quality of many studies" (p. 126), but he recommends that video recording should be used along with other data collection approaches and not instead of them. It can be used to supplement observations, interviews, and other data collection methods and be a valuable tool for improving the quality of the study. Videotaping lessons can produce a record that can be used to make very detailed transcripts of what occurred, and these can be replayed over and over again to pick up details and ensure accuracy, which would not be possible with field notes of the same lesson. So video recordings provide a way to fill in field-note records with details of such things as facial expression, emotions, and other instances of nonverbal communication that can easily be missed in field-note records (Hatch, 2002). Graue and Walsh (1998) point out limitations of using video recordings: for example, the time factor (i.e., the time required to view them and write them up if one intends to do so). A second limitation is the illusion of "being there" because watching video can give the viewer a false sense of experience. This was not the case in this study because I *was* always there and if I was not videotaping someone else was doing that for me and I was taking field notes or interacting with the children. Graue and Walsh (1998) also give some good advice on using video cameras in classroom research, such as the importance of a warm-up period to satisfy the children's curiosity and let them get used to the camera. Another important point made is that when the camera is left running without the researcher attending to it (looking through the viewer) it is less obtrusive than when the researcher is operating it though, of course, insensitive to any interesting features outside of the field of view.

I usually put the video camera up at the back of the classroom and left it running during the lesson. I sometimes also took the camera, walked around, and videotaped children's interactions or conversations by pointing the camera towards a specific child or children or at the teacher. However, I tried my best to get as good a picture as possible of the whole classroom, the teaching, and the interactions that took place. On two occasions, another teacher came and videotaped for me. That helped greatly because then I could take field notes myself or just listen, watch, and interact with the children and the teacher while the other teacher was taping. But writing a lot down (taking field notes) in the classroom while the teacher is teaching could be even be more disrupting than videotaping, because people are often very curious about what it is you are writing about. According to Bogdan and Biklen (2003) it is important not to act as if you are writing secrets and not to walk around continually with pen and pencil in hand. Reiss (2000) in his longitudinal study functioned as a nonparticipant observer during his study, sitting quietly at the back of the classroom, taking detailed field notes. According to him, this he believed, "had comparatively little effect on what went on in the lessons. Pupils, in particular, came to regard me as "part of the furniture" (p. 7). Sitting at the back, however, could have some disadvantage in that he was more likely to hear and see only what the pupils close to him were doing.

Looking at the advantages of videotaping over audio taping there are several points, but it depends on the purpose of the observation and the information the researcher wants to record. I can go over a videotaped lesson again, trace who said what and under what circumstances. Audio taping does not give the same opportunities although it can be less obtrusive and sometimes more flexible, especially, for example, when recording an interesting conversation or when talking to particular children (Graue & Walsh, 1998). A significant advantage of the videotapes in my case was that they give the opportunity to write up very detailed information afterwards and also to review the lessons at any time. The disadvantages are that at the beginning of the research the camera attracted much attention which caused disruption, but this did not last long as I did what Graue and Walsh (1998) described as a warm-up and introduced the camera and the purpose of it for the children and explained to them what I was doing. The children became used to the camera very soon and did not bother about it. Another disadvantage that I noticed was that while videotaping some children or activity, I might not notice what other children were saying or doing. Choosing what to videotape is therefore important. Having someone to undertake the videotaping while the researcher is taking notes seems ideal, but that "someone" is usually not available and, anyway, has the disadvantage that the one who is taping has not the same feeling for the subject as the researcher and may videotape things not so relevant to the research.

To a certain extent some of my classroom observation was participant observation where I took part in an activity or discussion, but this was not always the case. I did not decide beforehand how much I would participate, as Graue and Walsh (1998) suggest researchers should do. In some lessons I just observed and was usually behind or beside the camera at the back of the classroom. In other lessons I walked around and looked at children's activities and drawings. I sometimes helped the children write their names and the date on their drawings and sometimes asked what they were doing or what they were drawing. However, I never took over the teaching, although I sometimes added a word or two or asked an additional question to the class. When the class was divided into groups and the children were doing different activities, I sometimes helped one group while the teacher was helping another one.

Exactly how much a researcher participates can vary during the course of a study (Bogdan & Bilken, 2003). That was the case in this research and because I used to be a primary school teacher, I found it was sometimes difficult not to add a word or two. Not that I thought that the teacher was not doing well; it was just so tempting and I sometimes could not resist! One thing that has to be borne in mind is that data might be missed while the researcher is participating rather than taking field notes or videotaping and it can be problematic if participant observers give up important parts of their observer role to the participant role (Hatch, 2002). I did not decide beforehand what I was going to videotape as Graue and Walsh (1998) also suggest researcher should do. The video was usually situated in a place where it could catch as complete a picture as possible and the main aim was to get as good a picture as possible of the discussion, the interactions, the instructions, the activities, and the resources that were used.

After the lessons, I watched the video and transcribed in detail both what was said and also what was happening. It was very time consuming and took many days to transcribe all these video recordings. I printed each transcript out and read through it carefully, underlined words and phrases (coded the text), looked for patterns (key features), and identified themes. I wrote memos and summarized what I had discovered and tried to develop explanations and interpret what I had learned and seen. I also used information from classroom observation when cross analyzing the different sources of data that I had, namely, from interviews, drawings, and diagnostic tasks. I did open coding where I underlined single words and phrases in my field notes. I also used axial coding which involves assembling the data in new ways after open coding (Robson 2002), trying to relate categories to their subcategories in order to get more focused and more precise explanations on the issues or phenomena. Corbin and Strauss (1996) and Emerson et al. (1995) stress that the researcher using grounded theory should make observations, write field notes, code the notes in analytic categories, and develop theoretical proposition, which makes analysis both inductive and deductive. The third stage of coding (after open coding and axial coding) is selective coding. There, one aspect is selected as the core category and focused on. The basis for doing this arises from axial coding which gives a picture of the relationship between the different categories (Robson, 2002).

Early in the study both the teacher and I talked about those children who had been active in the lesson and those who had not been active. By being active we both meant taking part in the discussion or an activity. When we talked about being "not active" we meant passive, quiet, and not taking initiative in anything that had to do with educational activities in the classroom. We both did this even though, at the same time, we saw and knew that being active can also mean a lot more than the actions that we can see, which obviously calls for a much more stringent or more elaborate set of criteria for activity because a child can be an active listener and take in things that he or she sees without being "visibly" active.

The criterion used here for "classroom discussion" is when approximately 1/3rd of the class share their ideas with the teacher and the rest of the class. Classroom discussion is not a dialogue between the teacher and one child.

When the teacher and I talked about the children being active we used the words in a very simple way, especially at the beginning of the study. We talked about those who had been active in the lesson and those who had not been active at all. The criteria we used as the basis for our judgement was mainly verbal activities, that is, taking part in discussion or sharing ideas. I had difficulties in learning the names of a few children because they never said a word and never expressed themselves or took part in classroom discussion. I soon realized that there were about four children (five in Primary 2) that took a particularly active part in discussion. In the light of this, an eight level quietness scale (1 = most active, 8 = most quiet) was made in order to see if there were any differences between the development of the ideas of the children that took an active part in discussion and the more quiet ones.

According to the scale, children were divided into three groups: The *visibly active group*, the *semi-active group*, and the *visibly passive group*. The visibly active group consists of the children who took an active part in discussion. The children that hardly ever said a word constitute the visibly passive group and the rest, the children who were in between made up the semi-active group. The teacher and I grouped the children independently into these three groups on the basis of our overall impression of the children's participation in class discussion. In the three cases where we did not originally classify them in the same group we came to a joint conclusion after a discussion. Subsequently, I placed them on the 8-level quietness scale again consulting the teacher. The mapping between the scale and the groups was such that the visibly active group corresponds to levels 1–3, the semi active group corresponds to levels 4–6, and the visibly passive group to levels 7–8.

3.7   Observation of Teachers' Meetings

I sat in on six meetings with the four teachers when they were planning the lessons about the body each meeting took roughly an hour. Usually I came after the teachers had discussed and planned other things than the project about the body or I came at the beginning of the meeting and left when they had finished talking about the body project. I observed the meetings and took field notes. Before the first meeting, I gave the teachers a photocopy of a chapter in the book *Living Processes*, where there is discussion about children's ideas about the human body (Black & Harlen, 1995) and asked them to read the chapter so as to prepare them for the project. The teachers decided to divide the project about the body into three main themes or episodes in the first year according to the *National Curriculum Guide for Compulsory School: Natural Science* (Primary 1):

1. What do we know about the body (external body parts, organs, bones, muscles, joints)?
2. The heart is a muscle that pumps blood.
3. Healthy living—food and the food in the stomach

During the second year (Primary 2), there were five main themes or episodes:

1. Reproduction
2. The skin, the nerve cells, and the senses
3. The heart, the lungs, and blood circulation
4. The stomach, digestion, (liver and kidneys)
5. The brain

The teachers' meetings were held in one of the classrooms, not always the same one. I sat at the table with them, listened, took notes, and also participated in the discussion, pointing out things and coming up with new ideas. After each meeting, I wrote some main points up from my notes and sent these to the teachers by e-mail as an attachment. My points were usually about the planning of the project, what was needed, and organization of the class such as group work or hands-on activities. No one asked me to do this: It just seemed appropriate. Although I came up with some of the ideas and took part in the discussion, the teaching methods implemented were more or less their own ideas. I usually summed up the discussion before I left the meeting in order to clarify things for them and for myself. I tried to take down detailed field notes because I did not videotape or audio tape the meetings. I wrote the "report" on my computer afterwards. Writing down a lot at a teacher meeting was different from doing this in the classroom: It was not disruptive at all but was expected and was more like taking notes in order to write the minutes of a meeting.

Classroom observation was quite different from observation of the teachers' meetings. Observation of the meetings was much more participant. I often saw myself as one of the teachers in the group and I felt that they expected me to join in and I felt free to come up with ideas. Given that I am one of the authors of the teaching material *Let's Look at the Body* (Óskarsdóttir & Hermannsdóttir, 2001a and b) that they were using, the teachers naturally expected me to contribute so I sometimes also felt like an advisor helping them clarify things, by giving advice and providing ideas.

Sometimes I wondered if being the author of the teaching material was a problem—if the teachers perhaps felt they had to teach it as recommended in the teaching guidelines or if they planned it as they thought I would like them to do. However, they came up with and used many of their own ideas. I also thought that because I had been an adviser in the school on another project the teachers might still look on me as an advisor although the teachers knew that my role now was different.

The analysis of these teachers' meetings was rather different from analyzing classroom observations and not as difficult and time consuming. I read all the notes and transcripts through and identified key features and made comments, codes, and summaries.

3.8 Interviews With the Teacher

Interviewing is probably the most commonly used qualitative method (Delamont, 2002; Merriam, 1988; Yin, 2003). An interview is a conversation, usually between two people, but where one of them, the interviewer, is seeking information and responses from the other, the interviewee (Gillham, 2000). The form and the style of an interview depends on its purpose. The purpose of classroom or educational research is to obtain information on issues relevant to the general aims and questions of a research project (Gillham, 2000). Three main types of research interviews can be identified in qualitative research according to Delamont (2002). The first is the "interview" that is done during observation, when the researcher puts a quick question to the informant(s) about what is happening or to clarify something. Then there is the more formal interview where a checklist of issues is covered. This interview is more likely to be tape recorded. Finally, there is the life history interview, which can take many hours or even be spread over a period of years (Delamont, 2002). Those used in this study were the first two. While observing the lesson, I often asked the teacher a question or two in order to clarify or get information, but mostly I used the second type described here—a short or a long interview with the teacher where a checklist had been made beforehand. These so called qualitative interviews are often called semi-structured interviews and the researcher asks mainly open-ended questions (Hatch, 2002). It can also be described as a guided conversation where the goal is to elicit from the interviewee detailed data that can be used in qualitative analysis (Lofland & Lofland, 1995).

At the end of each lesson or topic, a short interview with the teacher was undertaken in the classroom or in the teacher tea room, just taking five to ten minutes. I either videotaped or audiotaped these interviews and wrote them up later. There was no particular purpose in videotaping (beyond audiotaping) but I happened to be videotaping a lesson anyway and so it was convenient, allowing me to use the audiotape record when transcribing the interviews later. In these interviews, the focus was on the lesson or the topic and the children's reactions. In short, the teacher was asked to evaluate the lesson, her own teaching, the materials she was using, and the reactions of the children. These short interviews were not preplanned in the sense that I had not written any questions to ask, the questions just came during the lesson and one thing led to another.

I also undertook three open extensive interviews with the teacher. According to Lofland and Lofland (1995) the "intensive" interview seeks to "discover the informant's experience of a particular topic or situation" (p. 18). The first extensive interview was undertaken at the teacher's home just one month after the research started where we looked through some of the videos and discussed some important issues that emerged from them. I thought this was good to help me understand the setting and how she organized things, and also to get to "know" the children better. The questions asked and the discussion that took place was not preplanned but when either of us saw or heard something on the video that needed to be explored, discussed, or explained further, we simply did so; and we often stopped the video because of our discussion and replayed some parts of it for further information. This first extensive interview was time consuming, taking between two and three hours, but I found it important, as did the teacher. The second extensive interview took place in the classroom near the end of the project. This took an hour. Beforehand a semi-structured interview schedule had been made with questions and discussion points that I wanted to explore (see Appendix I). I tried not to ask the questions directly, but made sure all the issues were covered and got her ideas and views on these issues. The third and last extensive interview was undertaken in a café in Reykjavik a few days after fieldwork had ended. It was really to follow up some important issues that had emerged from the other interviews and to tie up some loose ends. It also took about an hour.

Several different methods were used to analyze the interviews. As discussed, discourse analysis was used to give a clearer picture and understanding of how certain words and concepts were used and understood (Banister & Parker, 1994; Lofland & Lofland, 1995). I used some elements from grounded theory to identify themes in my interview data (Corbin & Strauss, 1996) as described above and I also used elements from a hermeneutical approach to obtain an understanding of the opinions, issues, and ideas expressed in the interviews in order to be able to interpret the findings (Gustavson, 1996; Hatch, 2002; Kvale, 1996). When I had completed each interview with the teacher I transcribed it; printed the transcript out; read carefully through it and underlined words, phrases, and concepts; wrote memos in between the text; and produced a summary of each transcript. When a concept or an idea occurred to me, I wrote it into the transcript as an analytic note and on these bits and pieces of analysis or memos and codes, I built a larger analysis (Lofland & Lofland, 1995).

3.9   Interviews With the Children

The purpose of interviews with children (and other people) is to get them to talk about what they know. However, children are different from adults and this makes them special interviewees and requires special methods. A lot of the literature on interviews with children is concerned with research within social-work research and is on difficult issues such as childcare, foster homes and other delicate issues (Hill, 1997), and also, for example, on children and family life, healthy eating, and subjects relating to children's daily life (Mauthner, 1997). However, when interviewing children, there are certain issues that are important to have in mind: For example, can the researcher let the children have a greater say in setting the terms of the conversation (Mauthner, 1997)? Individual interviews are usually more personalized and private than a group discussion and it is often easier for the child to be open in a one-to-one situation. On the other hand, talking to a strange interviewer on one's own can be stressful and inhibiting for a child. Group interviews enable children to interact and comment on each other's contributions and it can also be stimulating for individual children to be reminded of things they might otherwise not have mentioned. However, a disadvantage of group discussion can be that certain individuals may dominate the discussion (Hill, 1997). Mauthner (1997) argues for a child-centered approach to data collection, which views children as subjects rather than objects, meaning that there are important things to bear in mind in all research with children: first, let the child have a greater say in setting the terms of the conversation; secondly, the researcher should draw out children's subjective experiences to encourage them to describe events and issues through storytelling and anecdotes; and thirdly, the researcher can also consider children's experiences of the research process in the video- or audio-tape recording (Mauthner, 1997).

Individual interviews are usually more appropriate with older children since young children, five to six years old, can find them awkward and not an enjoyable activity. The younger the children the more difficult individual interviews can be. Young children frequently either remain silent or answer by using monosyllables or by just saying "I don't know." The researcher needs to find the balance between asking too many questions and too few but in practice it can be difficult to keep questions to a minimum when trying to cover what you want to get from the interview. The language used in the interview is another issue that needs attention: it is generally more profitable to encourage children to use their own language and their own way of communication and ask them to clarify when necessary (Graue & Walsh, 1998; Mauthner, 1997; Williams, Wetton, & Moon, 1989). Timing can also be an important factor: Interviewing children when there are more interesting activities available to them is not a good idea, so it is wise to select a time when the researcher is not taking anything important from the child.

When it comes to recording the interview, there are several points to bear in mind. Recording interviews on audio tape or videotape is common. Graue and Walsh refer to their own experience and say they believe that children are not as shy talking freely into a recorder as adults are and they favor the video recorder over the audiotape. "Viewing a video brings me back into the interview in a way that audio does not" (Graue & Walsh, 1998, p. 117). Video records children's facial expressions and the expressions often give more information than words. Using video, however, can be very obtrusive and it requires the interviewer to position people and take light and shadows into account. To get young children to focus on an activity during an interview can be useful, for example, to structure the interview around a specific activity, or to get them to draw and/or play games, which can make them focus and talk more easily (Mauthner, 1997).

At the end of the project, but before they completed the diagnostic tasks, all nineteen children were interviewed individually. I had made beforehand a list of questions or rather issues (see Appendix II) in order to see if they identified and knew the names of certain organs. I also wanted to know if they knew the location, purpose, and function of these organs and if and how they were connected. I also wanted them to show me how food went through the body from mouth to anus. I thought it was important to get information from each individual child in an interview so that information got from the interview could be cross checked with their drawings and the diagnostic tests they completed a little later. The interviews took place in one of the working rooms for the teachers at the school. A big torso of the upper half of the human body was used in the interview as I thought it would be easier to get them to talk and express their ideas if they had something visual to refer to and manipulate and I also thought this was a good way of getting them involved. It became like a little game or a puzzle when they were trying to put the organs back into the right places of the torso. Before each child entered the room, I had removed some organs from the torso and put them on the table. Then the child was asked about the organs (one organ at a time) and encouraged to show the way the food goes from the mouth and through the body.

Their own drawings were used in the interview. I showed them the drawings they had made early in the project and also those they made later and tried to get them to tell me what they had learned. They were also asked if they had books about the body at home. All the interviews were videotaped. The video camera was put on a stand and I made sure it gave a good view of the child and the torso (see Mautner, 1997). Each interview lasted for about twenty minutes and it took me three days to interview all the children. They were all very keen on coming to be interviewed. I tried to make a nice and unthreatening atmosphere but despite that, some of the children unfortunately did not express themselves much, merely answering my questions with one or two words and being very shy, which was also reported in other research (see Graue & Walsh, 1998; Mauthner, 1997; Williams et al., 1989).

Analyzing the children's videos was quite complex, having in mind all the facial and bodily expressions, the discussion, and what they did, pointing at and putting organs in their places in the torso. I watched all the interviews and transcribed them almost word for word and also wrote down descriptions of visual things on the video. Here, I also used all the different methods I have already described, underlining words and phrases, "coding," and writing memos. I also made a form where I wrote notes of their knowledge and understanding of issues such as location, structure, function, and processes, on the basis of the interviews (Appendix III).

3.10   Interviews With Parents

A sample of parents was interviewed. I chose parents of two of the children but asked the teacher to pick the others by random (parents of two girls and two boys). I chose the parents of these two children, a boy and a girl, because they had contributed more than the others to the research by expressing their ideas and were very interested in the project. I believed parents of six children would be appropriate for the purpose of this part of the research, to gain access to information about the child's interest in the human body and resources available in the home. I phoned personally the parent/parents of each child, and asked if I could meet them and talk to them about issues involved in my doctoral research about children's ideas about the human body. They all seemed familiar with the project from school and were all happy to take part. I decided that it would be best to visit them in their homes because then they would feel more relaxed than coming, for example, to the university. They all agreed with this except for one who worked in the neighborhood of my university and asked if she could come to me, because it would be more convenient for her. The parental interviews tended to be more formal than the interviews with the teacher (see Appendix IV). The parents that I talked to were strangers to me and I found that I had to make a special effort to prepare for the interviews (see Delamont, 2002). I made a checklist of issues and questions to cover and though I did not ask the questions directly from the paper, I looked over it at the end of the interview to see if I had forgotten anything. In one case both parents were present but in the other five cases only the mother. The children were not present on any occasion but I had not mentioned anything about that to the parents. However, I saw some of them playing in their room or outside their home and they thought it was very exciting that I was coming to their home and one asked me to come again soon. The interviews with the parents lasted from 20 minutes to an hour. I tape-recorded all the interviews (asking the parents for permission beforehand), using a very small tape recorder or an MP3 player.

Analyzing the parents' interviews was not as time consuming as analyzing the teacher interviews as they were more formal and not so long. Here I did not transcribe all of each interview but merely listened to the audiotape and wrote down issues relevant to the study as described before.

3.11 Drawings

According to Greig and Taylor (1999): "Children's drawings are believed to reveal the child's inner mind" (p. 117). As drawings are believed to be an important way of getting access to children's ideas about various things, I decided from the beginning of my research to use children's drawings as a source of information. As a former primary school teacher I knew that young children usually liked to draw and often expressed themselves through drawings, so I was convinced I could get important data from them using this method.

A number of studies have been done on the use of drawings to probe children's ideas about their bodies and have proved to be an important and a valuable way of getting information about children's ideas about various things (Black & Harlen, 1995; Carvalho, Silva, Lima, & Coquet, 2004; Cuthbert, 2000; Guichard, 1995; Haney, Russel, & Bebell, 2004; Osborne, Wadsworth, & Black, 1992; Reiss & Tunnicliffe, 1999a, 1999b; Reiss et al., 2002; Teixeira, 2000; Tunnicliffe, 2004; Tunnicliffe & Reiss, 1998, 1999a). According to Haney et al. (2004) drawings can be of great value for assessment and also for evaluation of educational practices and issues, and "students" drawings provide a rich opportunity to document students' perspectives" (p. 15). They also point out that drawings have been neglected as a tool of educational research and say that when using drawings in research it is probably best to interview the artists about their drawing (Haney et al., 2004). This was done in this study but not until the interview at the end. According to Reiss and Tunnicliffe (2001) drawings are of special value for children that have difficulties in expressing themselves verbally. Drawings can also be helpful for those children that are shy in conversation or who lack certain linguistic skills or speak only a foreign language (Reiss & Tunnicliffe, 2001). There was one foreign boy in the class in my study who spoke hardly any Icelandic, so I saw this as an important way to get insight into his ideas. There were also some children in the class (about one third) that did not express themselves much verbally, so I thought this was an important way to get access to their ideas. Ten times, the children made drawings, see numbering in Table 3.1. I had chosen two pictures of the outlines of the body and the teacher photocopied them for the class. Template 3.1 was used when they drew the bones and Template 3.2, slightly different, where they drew the organs.

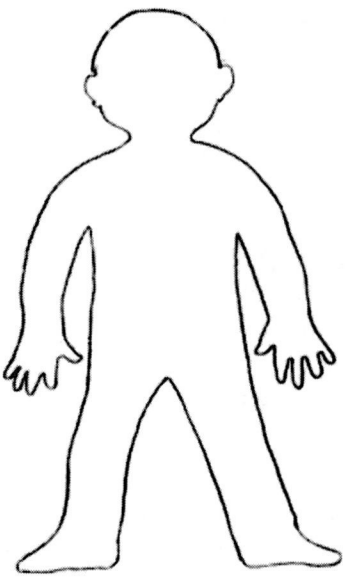

Template 3.1

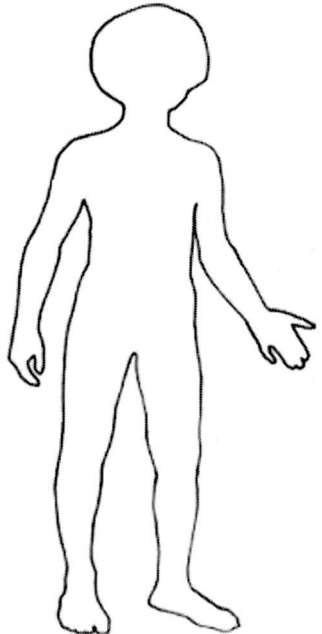

Template 3.2

Different templates were used so it would be easier to classify the pictures and analyze them. The children were asked to draw bones and organs into a previously made outline of the body so they would concentrate only on what was "inside" the body, otherwise some might put much effort into drawing the outlines, hair, and eyes and forget the things they were asked to do. In addition, children's capabilities to draw are very different, so having a previously made outline gave them to some extent the same starting point. The children also made a free hand drawing of a model of the heart, they were given a photocopy of the human skeleton (see Template 3.3) where they had to draw the muscles on the bones. They also drew on a previously made picture of a part of the digestive system where they had to show the food in the mouth and in the stomach (see Template 3.4).

The children had as much time as they needed for the drawings but usually it did not take them more than 5–10 minutes to finish each drawing. All the drawings were collected and classified using the date and name of the child, so I could see if and how their ideas developed and changed. I also used the drawings in the interviews with the children where we looked at them together and I asked them to comment on or explain certain issues.

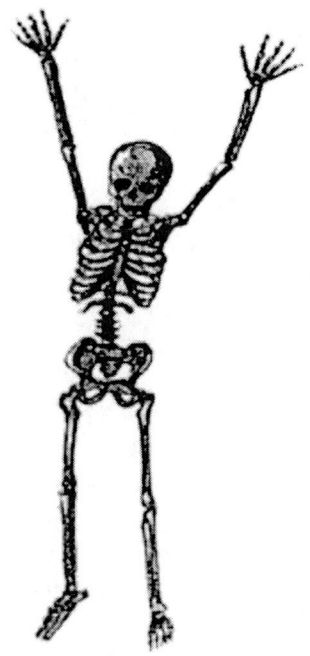

Template 3.3

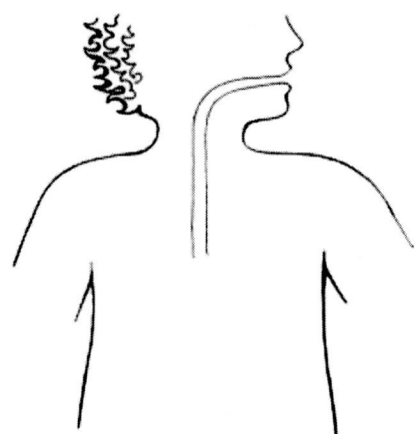

Template 3.4 Graphic by Ragnheiður Gestsdóttir.

Children's drawings have been used to assess children's knowledge. An extensive study, the English Primary SPACE Project (*Science Processes and Concepts Exploration*), discussed in the literature review, used a range of methods, including drawings, to assess children's knowledge of a variety of things in their environment, including their body (Harlen, 1992; Osborne et al., 1992). Drawings were also used in analyzing human figure drawings made by 8–11 years old in the UK and Japan (Cox, Perara, Koyasu, & Hiranuma, 2001). These drawings were rated and categorized according to how the face, arms, and legs were drawn, and were compared with the findings of previous research studies.

Drawings were also used in a study of children aged 7 to 11 in the UK where the aim was to assess whether the children saw their body maps as integrated systems (Cuthbert, 2000). In a Portuguese study, drawings were used to obtain information about children's conceptions of digestion where the pupils were asked to draw what happened to a cookie they ate like mentioned in Chapter 2. The drawings were classified and analyzed in order to establish categories in relation to body structure or cookie digestion. Each drawing was inspected and then placed in an appropriate category (Carvalho et al., 2004).

Reiss et al. (2002) also did an extensive international study on children's ideas about the human body. They used a variety of approaches, especially drawings, to establish children's ideas about the structure and the place of the bones and different organs, and put forward definitions of the organ systems (Figure 3.1).

They developed a *seven level scale* reflecting different levels of biological understanding about the human skeleton, and another scale also with

Skeletal system	Skull, spine, ribs and limps
Gaseous exchange system	Two lungs, two bronchi, windpipe which joins to mouth and/or nose.
Nervous system	Brain, spinal cord, some peripheral nerve (e.g., optic nerve).
Digestive system	Through tube from mouth to anus and indication of convolutions and/or compartmentalisation.
Endocrine system	Two endocrine organs (e.g., thyroid, adrenals, pituitary) other than pancreas (scored within digestive system) or gonads (scored within urinogenital system).
Urinogenital system	Two kidneys, two ureters, bladder and urethra or two ovaries, two fallopian tubes and uterus or two testes, two epididymes and penis.
Muscular system	Two muscle groups (e.g., lower arm and thigh) with attached points of orgin.
Circulatory system	Heart, arteries and veins into and/or leaving heart and at least to some extent, all around the body.

Figure 3.1 Reiss and Tunnicliffe's definitions of the organ systems. *Source:* Reiss et al., 2002.

seven levels that reflected different levels of biological understanding about human organs and organ systems (Figure 3.2.) (Reiss & Tunnicliffe, 1999a, 2001; Reiss et al., 2002; Tunnicliffe & Reiss, 1999a).

A "portfolio" was made for each child with all their drawings in the time sequence to see if and how the drawings had changed over time. Another form was made to analyze the drawings (see Appendix V), and there I put information as to their knowledge and understanding of issues such as the place and the structure of the heart, lungs, stomach, small and large intestines, colon, and the brain; the function of the heart, lungs, and the brain; and the process of the food going through the digestive system. I went through all the drawings made by each child and marked and wrote down information on the form according to the information got from the drawings. The seven level scales developed by Reiss and Tunnicliffe were used when analyzing the drawings of the bones/skeletons and the organs. I, however, soon realized that their scale for the organs did not quite suit the drawings in this study as their scale was developed for a broader age group, not for a group of such young children. Their "organ" scale has bones as well as other organs in the same scale but for the sake of this research it was important to gather information about children's ideas about the bones and other organs separately. As the children in the study were only 6 years

Bones—Skeleton	
Level 1	No bones.
Level 2	Bones indicated by simple lines or circles.
Level 3	Bones indicated by 'dog bone shape' and at random or throughout body.
Level 4	One type of bone in its appropriate position.
Level 5	At least two types of bone (e.g., backbone and ribs) indicated in their appropriate position.
Level 6	Definite vertebrate skeletal organisation shown (i.e., backbone, skull and limbs and/or ribs).
Level 7	Comprehensive skeleton (i.e., connections between backbone, skull, limbs and ribs).
Organs	
Level 1	No representation of internal structure.
Level 2	One or more internal organs (e.g., bones and blood) placed at random.
Level 3	One internal organ (e.g., brain or heart) in appropriate position.
Level 4	Two or more internal organs (e.g., stomach and a bone 'unit' such as the ribs) in appropriate positions but no extensive relationships indicated between them.
Level 5	One organ system indicated (e.g., gut connecting head to anus or connections between heart and blood vessels).
Level 6	Two or three major organ systems indicated out of skeletal, gaseous exchange, nervous, digestive, reproductive, excretory, muscular and circulatory.
Level 7	Comprehensive representation with four or more organ systems indicated out of skeletal, gaseous exchange, nervous, digestive, reproductive, excretory, muscular and circulatory.

Figure 3.2 Reiss and Tunnicliffe's levels of biological understanding.

old, I also thought it would be easier for them to concentrate on one issue at a time in their drawing. In light of these observations, Reiss and Tunnicliffe's scale was used as a basis to develop a modified scale for the drawings of the organs (see Figure 3.3).

Level 1	No representation of internal structure.
Level 2	One internal organ (e.g., brain or heart) placed at random.
Level 3	One internal organ (e.g., brain or heart) in appropriate position.
Level 4	Two internal organs (e.g., brain, heart or stomach) placed at random.
Level 5	Two internal organs (e.g., brain, heart or stomach) in appropriate positions but no extensive relationships indicated between them.
Level 6	More than two internal organs in appropriate position but no extensive relationships indicated between them.
Level 7	More than two internal organs in appropriate position and one organ system indicated (e.g., gut connecting head to anus or connections between heart and blood vessels).
Level 8	Two or more major organ systems indicated out of digestive, circulatory, gaseous exchange and nervous systems.

Figure 3.3 Reiss and Tunnicliffe Scale of the organs as modified by the author, called here the modified R/T Scale.

3.12 Diagnostic Tasks

In order to get more information about the children's ideas and from a source other than classroom observation, drawings, and interviews, a number of tasks were designed as a means of discerning in a different way what pupils knew or did not know. As these were intended to probe the picture obtained by the other methods (i.e., drawings, observations, and interviews), these tasks are classified as "diagnostic tasks." They were intended to allow a diagnosis of the inferences deduced from the other tasks. So at the very end of the project the children were asked to complete a few written tasks, here called the diagnostic tasks (see Chapter 3). It is important to look across all the work relevant to a particular objective and not just judge from one. Each piece of work undertaken by the pupil and each observation made by the teacher can be used to build up a picture and demonstrate understanding of an idea (Harlen, 2000).

Some of the children in the study were shy and did not express themselves, either in class discussion or in the interview, and some of the drawings did not give a clear picture of their knowledge and understanding. Therefore, some diagnostic tasks were designed and shown to the teacher for her opinion. She agreed this would be valuable and give a broader picture of the abilities of each child. The children were asked to complete these tasks individually. This came at the very end of the project. The tasks aimed at referring to and covering all the issues that had been focused on during the project, and in some cases children were asked to complete a number of similar tasks, but using different methods (writing, completing, colouring, or pairing). The tasks were in two separate "booklets" and they had to complete one before they were handed the other (see tasks in Appendix VI).

In the research literature there is relatively little on how diagnostic tasks can be used to get access to children's ideas, as such tasks are more often used for summative purposes as a subsequent or follow up to learning or as an end of course test (White & Gunstone, 1992). Diagnostic assessment has a specific focus, according to Hollins, Whitby, Lander, Parson, & Williams. (1998), as it is concerned with examining a particular area of performance and is some-times assessed through special tests whereas summative assessment involves the summing up of where a child has reached at the end of a particular time and provides a picture of a child's development at the end of a year, term, or key stage (Hollins et al., 1998). Harlen (1993), however, talks about on-going assessment as "formative" or "diagnostic" where its purpose is:

> to help teachers in making decisions about learning experiences of indi-vidual pupils and of a group of pupils. It is part of day-to-day planning and teaching. It involves identification of where children are in their learning to inform the action to take. (p. 138)

She also discusses assessment for summative purposes that gives information on what children have achieved at certain times, which can be collected, for example, at the end of a unit of work, designed especially to assess the point reached in their development of ideas (Harlen, 2000). According to Harlen, summative assessment is not intended to guide learning and is assessment of learning rather than assessment for learning. Farmery (2002) talks about both formative and diagnostic assessment whereas both can be used to in-form the next stage of learning in science. Formative assessment is the term for ongoing, everyday assessment, which is used for building upon and for deciding the next steps to be taken. Diagnostic assessment, however, helps the teacher analyze and classify learning difficulties, so that appropriate as-sistance and interventions can take place (Farmery, 2002). I did not have this especially in mind when I decided to ask the children to complete the diag-nostic tasks. In the case study about "Ourselves" (Frost, 1997), different tasks and methods were used in monitoring overall learning and children's work was analyzed to see what the children knew and could do. This was what I had in mind; that is, to get information from different sources.

According to Farmery (2005), assessment tasks that involve practical activities or investigations are important, along with other types of activities like questions that the children have to answer. These kinds of tasks are designed for the pupils to demonstrate their skills, knowledge, understanding, and attitudes that are open-ended and open to interpretation by the pupils. They are also useful for providing assessment evidence, as they are completed individually and therefore give a summative assessment of the pupils' learning (Farmery, 2005). These tasks can be used individually and also with pairs or groups of pupils, which Farmery says are very important for pupils who are reluctant to engage in activities unaided or to carry out a one-to-one discussion with the teacher. She also talks about formal tests at the end of topics, which can be used formatively for the pupils, in that they are used to amend future planning but are designed to ascertain what the learner has learned. According to Farmery these can be

- closed questions requiring a one-word answer,
- ticking one of four statements,
- deciding whether a statement is true or false,
- short explanations of scientific models or theories, and
- identifying errors in statements or diagrams (p. 124).

These are the seven tasks that were used in the study under the framework of diagnostic tasks (Appendix VI).

I. One task involved drawing a line between a name of an organ and the function of it.
II. Another task included statements and the children were asked to write *T* (True) and *F* (False) accordingly.
III. The third task involved coloring organs in certain colors.
IV. In the fourth task the children were asked to draw with a red color the way the food travels from the mouth and through the body.
V. In the fifth task they were asked to connect the words to the appropriate spots on a drawing of the skeleton.
VI. The last two (VIa and b) were questions and answers tasks.

Day-to-day activities also provide teachers with information about children's ideas and learning. Frost (1997) mentions four activities which were planned especially to help the teacher get information about what the children knew about the human body and also what they could do: brainstorming exercises, problem solving exercises, drawing tasks, and an exercise in interpreting data, which was done systematically with each child (Frost, 1997).

When the children in the study had completed their diagnostic tasks, I collected all the scripts and put them in each child's portfolio. I used the same form for analyzing these tasks as I used for the drawings, inserting notes of their knowledge and understanding of issues such as the location, structure, function, and processes as indicated by their diagnostic tasks (Appendix VII).

It is important to gather information from different perspectives in order to get a more holistic view of children's ideas. Ward, Roden, Hewlett, and Foreman (2005) emphasize that there are only three ways of gaining evidence about how pupils are learning: by observation, by discussion, and by marking and looking at completed tasks. In reality, it can be a complicated task for the teacher to assess through marking written tasks, drawings, or responses, and this will still not provide a full picture of the processes and learning that have taken place. It is therefore important to use a mix of many different approaches but also to understand the advantages and disadvantages of each. With older children, for example, assessing written tasks can provide reasonable information, while with younger children the assessment of written tasks provides limited insight, particularly about the processes of learning (Ward et al., 2005).

Thus, the prime aim of all the data gathering and analysis in this study was to get information about the children's ideas. Different methods were used to obtain as full a picture as possible of these ideas and their development, but the variety was also in some instances to allow some validation of the data by a way of triangulation. Information emerging from the different tasks and through different methods of enquiry—classroom observations, interviews, drawings, and diagnostic tasks—were compared to discern a pattern in the children's ideas and assess the changes in their development during the course of the study. This approach also allows comparison of the different methods used to collect data (drawings, interviews, and diagnostic tasks).

3.13 Statistical Analysis

The exploratory and qualitative methodological approach adopted in this study, where an attempt is made to be faithful to the teaching situation, while stepping outside it in the role of the researcher, does not assume the quantitative approach often used in the educational research. However, when looking at change, or the relationship between different methods, this often invites or necessitates statistical inferences. Therefore, the data is frequently mapped onto scales as described earlier in the chapter and routine statistical analysis used, in particular the Student's *t*-test and Pearson correlation. In most cases the *t*-test used, is the paired test, and when progress is being investigated a one-tailed test is being used. When the p-level is less than 0.05, a difference is considered significant, but when the p level is lower than 0.01, this level is used.

4

Results

In this chapter, the findings from the research are presented. In Section 4.1, I will describe the ideas the children in the study had about the body before teaching about the body started and how the ideas changed. I looked for data on the development of ideas about structure, position, function, and processes of bones, organs, and organ systems. In order to do this I have to describe to a certain extent the teaching or intervention that took place even though it is described in more detail in Section 4.2 and consequently, there is some repetition. In Section 4.2, I describe the different teaching methods the teacher used when teaching about the body, and the teaching material used and how these influenced the changes in the children's ideas. In Section 4.3, I describe the interaction that took place, pupils' involvement, the difference between the involvement of boys and girls, and the influence the children have on the development of each other's ideas. Special focus will be put on the quiet and withdrawn children and their development of ideas.

For clarification, phrases that refer to the children having done or taken part in certain things, mean that all the children took part in all the

"The Brain Controls Everything", pages 97–154
Copyright © 2016 by Information Age Publishing
All rights of reproduction in any form reserved.

things unless otherwise indicated. The phrase, *the majority of children*, indicates 16 or more (of the 20/19 children). The *minority* is less than five. In other cases the actual number will be stated.

4.1   Children's Ideas About the Body and How Their Ideas Change

The children in the study bring to school and share the same or similar general ideas about the body. According to the classroom observation, the teacher started each topic by finding out children's ideas followed by some kind of intervention (demonstration, activities, drama, . . .) and concluding with summary and evaluation. In Primary 1 the emphasis, according to the *National Curriculum Guide: Natural Science* (Menntamálaráðuneytið, 1999), is on the external body parts, bones and muscles, teeth, and healthy living. In Primary 2 the focus is on the function of main organs. The children are also supposed to know about reproduction, that is, that babies are made by one cell from the father and one from the mother. It is also emphasized that an account should be taken of children's preconceptions and prior knowledge and experiences, and this should be taken further and used to build on further experiences and knowledge.

The song "Head, shoulders, knees, and toes," a song about the external body parts, was used as an "ignition" at the very beginning of the project and a big doll was used in the discussion about the different body parts. Children in Iceland know this song from an early age and during the first lessons about the body they sang this song several times with the teacher. The teacher asked questions about the body and the children came up with their ideas, which the teacher wrote down on a big white poster hanging on the wall. The initial question was: "What do we know about our bodies?" The children were aware of the names of their external body parts and their senses. They were also aware they had bones, blood, and veins, and some main organs (e.g., the heart, the lungs, the brain, and the stomach), but they were generally not aware of the word "organ" and no one mentioned liver or kidneys. Comments made were such as: "The heart pumps blood but we are dying if it pumps slowly" and "The brain controls everything." They were aware of the importance of healthy living such as eating healthy food and going early to bed to get enough sleep, and of the senses, "You can see with your eyes, hear with your ears, smell with your nose." They were also aware that you have to clean yourself and wash your hands after going to the toilet. One child mentioned the skin being very important because it kept everything inside: "If there was no skin, the blood would just run out of the body and the bones would not be in the right places."

4.1.1 Bones/Skeleton and Muscles

During the first lesson in the first curricular episode the children received a white paper with an outline of the body, Template 3.1, and the teacher asked them to draw the bones in the body. The drawings of the bones were analyzed using the *Seven Level Scale* developed by Reiss and Tunnicliffe (Reiss & Tunnicliffe, 1999a; Tunnicliffe & Reiss, 1999a). Level 1 signifies the lowest level and 7 the highest (Figure 3.2). Fourteen children scored at Level 2: Bones indicated by simple lines or circles. Four children scored at Level 3: Bones indicated by "dog bone shape" and at random or throughout body. One child scored at Level 4: One type of bone in its appropriate position and one child scored at Level 6: Definite vertebrate skeletal organization shown (i.e., backbone, skull and limps, and/or ribs). That child, a boy named Óli, also drew joints. He was the only child who did that.

During the discussion about the bones, one child, a boy named Árni, said that the skeleton held the body upright and without it we would just be a pile and he made himself fall on the floor. Árni said it was important to drink milk and eat cheese and Óli said it was because there was calcium in milk and cheese and calcium was important for the bones.

The discussion about the muscles showed that all the children were aware of muscles, especially in their arms; examples of comments made were: "Because you use them when you have to move houses" and "If you want to win the strongest man of the world championship title."

After the discussion, the children did many different activities on the skeleton and the muscles. They all participated and did all the activities. These were: a paper puzzle activity where the children had to put together pieces of the skeleton, a puzzle activity on the computer where the children in pairs had to make a skeleton by putting the different bones together, exploration of a model of a human-sized skeleton, and pictures of the skeleton from the school textbook, *Let's Look at the Body* (Óskarsdóttir & Hermannsdóttir, 2001a). They also explored their own bones by feeling them and did different activities involving the joints, bending and bowing, imitating different animals, and walking up and down stairs with bent and straight legs. They were also sent home with the SHIPS (School Home Investigation in Primary Science) activity, *Moving Feet*, where they were supposed to work with their parents on an activity about the muscles and joints (Solomon & Lee, 1991).

Two weeks later, after having been involved in all these activities, the children again received a white paper with an outline of the body (exactly the same as before) and the teacher asked them to draw as detailed a drawing of the skeleton as they could. The drawings were again analyzed using

Reiss and Tunnicliffe's scales (Reiss & Tunnicliffe, 1999a; Tunnicliffe & Reiss, 1999a). Not surprisingly, all the children, but one, scored higher after the teaching about the skeleton and all but three scored two levels or more higher than before (see Figure 4.1). Seven children scored now at Level 6: Definite vertebrate skeletal organization shown (i.e., backbone, skull and limbs, and/or ribs). All the children drew the ribs after the intervention and many of them also the backbone and the skull. Many of them (ten children) also drew the joints after the intervention (e.g., knees, elbow, and shoulders) like the examples of children's drawings in Figures 4.2 and 4.3

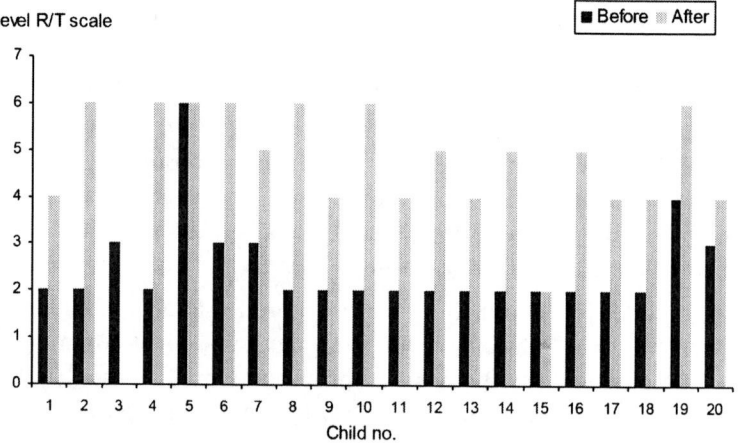

Figure 4.1 Ideas about the bones held by children P1 based on the drawings. The figure shows how the children scored on the Reiss and Tunnicliffe's scale before and after intervention. The average difference in children's performance is 2.37, $t(18) = 8.5$, $p < 0.01$, on a paired t-test.

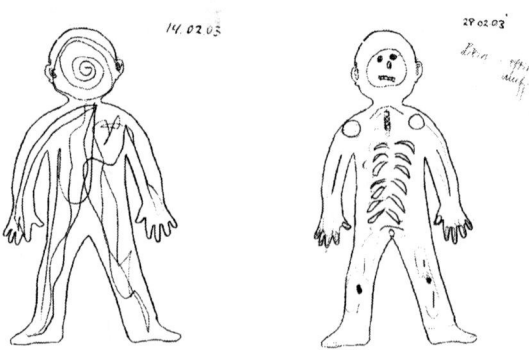

Figure 4.2 Drawings of bones before and after teaching from Child 1.

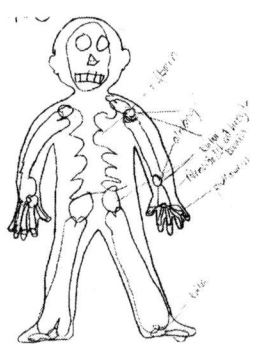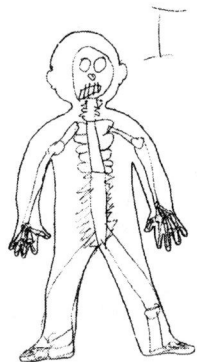

Figure 4.3 Drawings of bones before and after teaching from Child 5 (Óli.)

show. The child on Figure 4.2 scored at Level 2 on the drawing made before intervention, as he drew the bones as "simple lines or circles" and at Level 4 on the drawing made after intervention, "one type of bone in its appropriate position," here the ribs referring to the terminology in Reiss and Tunnicliffe's scale (Reiss & Tunnicliffe, 1999a; Tunnicliffe & Reiss, 1999a). The boy, Óli, scored at Level 6 before and after intervention, Figure 4.3. It can be seen that the later drawing is more detailed but it is not enough to move him up a level on Reiss and Tunnicliffe's scale (Reiss & Tunnicliffe, 1999a; Tunnicliffe & Reiss, 1999a).

In the interview at the end of the project fifteen children recognized the ribs on the torso; they also knew that they protected the heart, and some also knew that they protected the lungs and the stomach. Two of the four children that did not recognize the ribs were not in the class in Primary 1 and it is not known what kind of teaching they received in Primary 1 about the body or if any.

The children were asked to draw the muscles of the body on a drawing of the skeleton. Results from the drawings show that many of the children (thirteen) drew the muscles on the legs and on the upper arms; six children drew muscles elsewhere in the body although these children also drew the biggest muscles on the arms and legs (see Figure 4.4). In the diagnostic tests at the very end of the project all the children but one marked "true" in a true/false question: "The muscles help us to move the body."

Only one child in the class scored 100 percent on the diagnostic task where the children were asked to label all the bones of a skeleton with their right names. This child hardly understood any Icelandic at the beginning of the project one year earlier and had never taken part in the discussions.

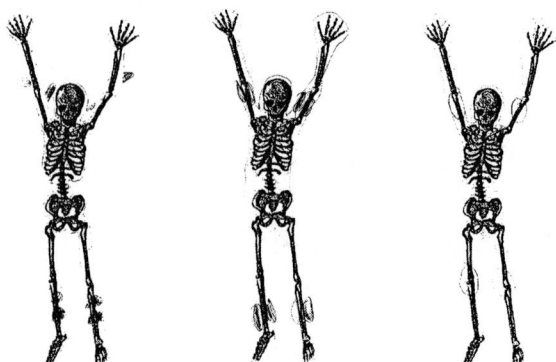

Figure 4.4   Three examples of children's drawings of muscles on Template 3.3.

The majority of the children did not know and could not label the shin and the collar bone.

The results from the drawings show that according to their initial drawings most of the children did not look at the bones in the body as a skeleton where all the bones are connected together, or at least they did not draw them that way. Most of them drew the bones as lines. They, however, had the idea that bones were all around the body. After the discussion and teaching about the bones and muscles, the majority (16 children) had moved up at least two levels on the Reiss and Tunnicliffe's scale (Reiss & Tunnicliffe, 1999a; Tunnicliffe & Reiss, 1999a) and all of them drew the ribs in an appropriate place. It is clear from their drawings that the activities they did by bending and stretching and walking up and down stairs with straight and bent legs had effect, and they understood the function and importance of joints and tendons after these activities. They also now seemed to have a picture in their mind of the skeleton as a frame that held the body upright, that is, they are aware of the structure and the function of the skeleton and drew the main bones and bone units (like the ribs) in relatively correct positions, even though they did not draw all the bones connected together.

According to comments made in the class discussion, the children believed the muscles were important if we have to do something physical like moving, carry heavy things, move a house, and to win the "strongest man in the world" championship. Most of them, however, drew the muscles only on the upper arms and legs even after so many different activities involving bending and stretching, and they did not seem to connect the muscles to the tendons, at least not on their drawings.

4.1.2 *The Organs in the Body*

At the beginning of the project the children were also given a white sheet of paper with an outline of the body (Template 3.2) where they were asked to draw what they thought was inside the body. This time they were asked not to draw the bones. The drawings were analyzed using the Seven Level Scale developed by Reiss and Tunnicliffe (Reiss & Tunnicliffe, 1999a; Tunnicliffe & Reiss, 1999a) and a form that I developed (see Appendix V). On the basis of their drawings, the results show that all twenty children drew a heart and seventeen drew a V-shaped "Valentine's" heart. Thirteen children drew veins all around inside the body. Three children, however, drew the body filled with blood. During a discussion before the children made this drawing, one boy had said that the skin was to keep all the blood inside the body otherwise it would all leak out. Another boy said that the heart was a muscle that was working all the time. There were also comments about the function of the heart like "We have veins and blood that run around inside the body," "We have a heart so we can live," and "The heart beats and pumps blood into our body." One said that we had lungs so we could breathe, but only four children drew lungs (or one lung) in their initial drawing.

The initial drawings were analyzed on Reiss and Tunnicliffe's Seven Level Scale (Reiss & Tunnicliffe, 1999a; Tunnicliffe & Reiss, 1999a). Eleven children scored at Level 4: Two or more internal organs (e.g., stomach and a bone "unit" such as the ribs) in appropriate positions but no extensive relationships indicated between them. The heart and the brain were the most frequently drawn organ. As mentioned before, only four children drew the lungs (or one lung) in their initial drawing and only six children drew the stomach or something that had to do with digestion. The rest did not draw the stomach or anything in connection with the digestive system.

One month later after interventions that included talking about the body, looking at pictures, doing different activities that mostly had to do with external body parts, and the bones, senses, and health issues, they were asked to draw again the organs that were inside their body. Before they were asked to do this drawing, the children had also done some interactive tasks on the Internet where they were asked to put the organs (i.e., the brain, heart, lungs, stomach, colon and intestine, and liver and kidneys), in their right places. These drawings showed that there was not nearly as much difference here between pre- and post-drawings as between the pre- and post-drawings of the bones. As shown on Figure 4.5, about half of the children (9 children) did not score higher than before, although many were drawing more organs than before, and as a result this development in their ideas cannot be seen in Figure 4.5

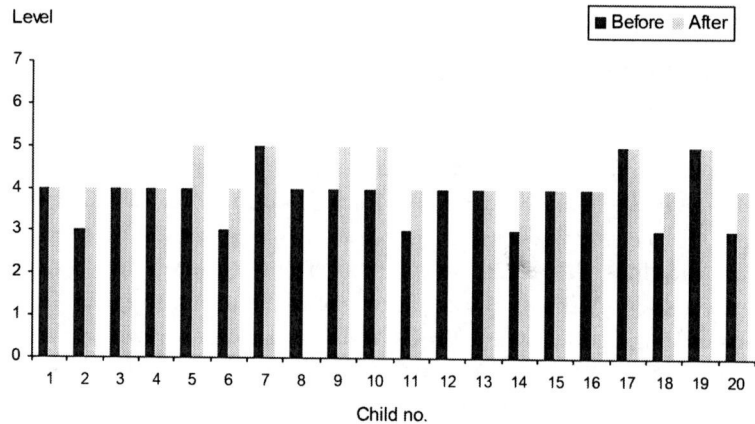

Figure 4.5 Ideas of organs held by children in P1 as measured by the R/T Scale. The average difference in the children's performance is 0.50, $t(17) = 4.1$, $p < 0.01$ on a paired *t*-test.

I found that Reiss and Tunnicliffe's scale (Reiss & Tunnicliffe, 1999a; Tunnicliffe & Reiss, 1999a) for analyzing the drawings of the organs did not reflect the change in ideas as Figure 4.5 shows very little variance, even though the average progress made by the children is significant. This is partly because they developed the scale for other purposes and for a wider age group, namely to see what children at different ages presented in a drawing about what they thought was inside their bodies. Therefore, I decided to modify the scale, as described in Chapter 3, so that the changes in ideas can be more easily seen, as shown in Figure 4.6.

According to my scale, six children did not move up a level after intervention, but two of these scored at Level 5, two at Level 6, and two of them at Level 7 in both pre- and post drawings. This shows that those children who scored at Level 6 and 7 already had substantial knowledge before intervention and the intervention did not (according to the drawings) add to that. However, twelve children moved up a level, two moved up three levels, and three moved up four levels, as seen on R/T modified scale.

As seen on Figure 4.7 the modified scale shows better the change in ideas that took place between pre- and post-drawings of the organs the Reiss and Tunnicliffe's scale (Reiss & Tunnicliffe, 1999a; Tunnicliffe & Reiss, 1999a).

Figure 4.8 shows Child 16 draws the brain and the heart in relatively correct position in the template used (Template 3.2) and scores therefore at Level 5. A month later the child has added the lungs and the kidneys

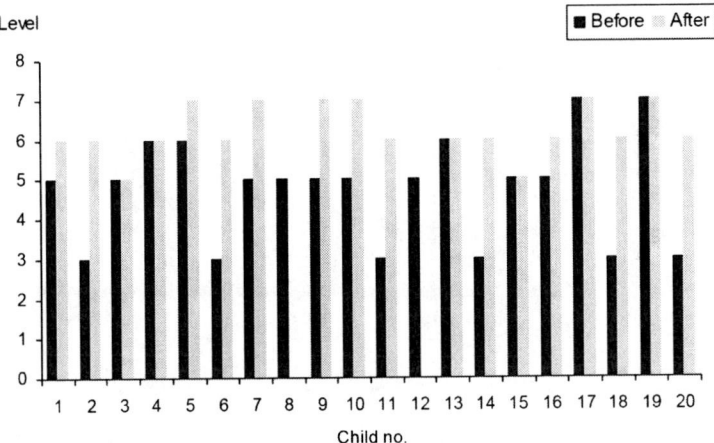

Figure 4.6 Ideas about the organs held by children in P1 as measured by the modified R/T Scale. The average difference in the children's performance is 1.50, $t(17) = 4.9$, $p < 0.01$ on a paired t-test.

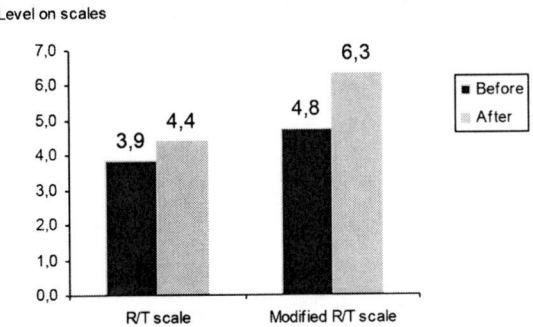

Figure 4.7 Difference on the performance of the children in P1 as measured by the R/T organ scale (on the left) and the modified R/T organ scale (on the right).

and although the location of these organs is not correct, the child scores at Level 6. Child 19, however, scores at Level 7 in the initial drawing as there are: More than two internal organs in appropriate position and one organ system indicated (e.g., gut connecting head to anus or connections between heart and blood vessels). Although the latter drawing is much more accurate and the child has added the liver to the drawing he or she is still at Level 7.

Figure 4.9 shows that, according to their post-drawings, the children are more aware of both the structure and the location of the heart and the brain, especially the brain, than of the structure and the location of

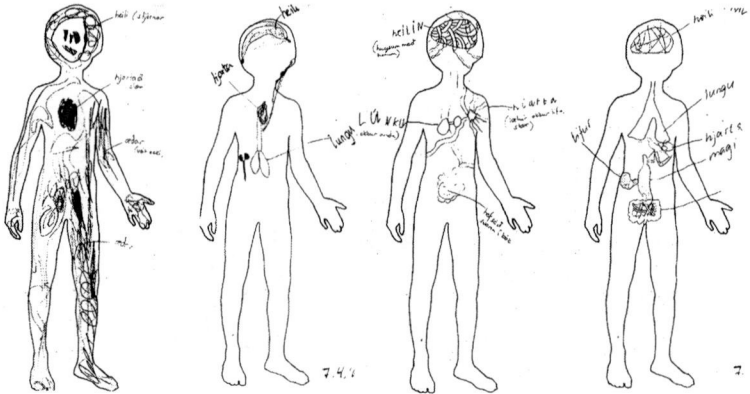

Figure 4.8 Two examples of children's drawings of the organs before and after intervention using the R/T modified scale.

the stomach and the lungs. However, they still represented the heart as V–shaped, but their drawings of the stomach and the lungs were not as accurate and the location was not clear in their mind. Data here are only from the seventeen children who were both in Primary 1 and 2.

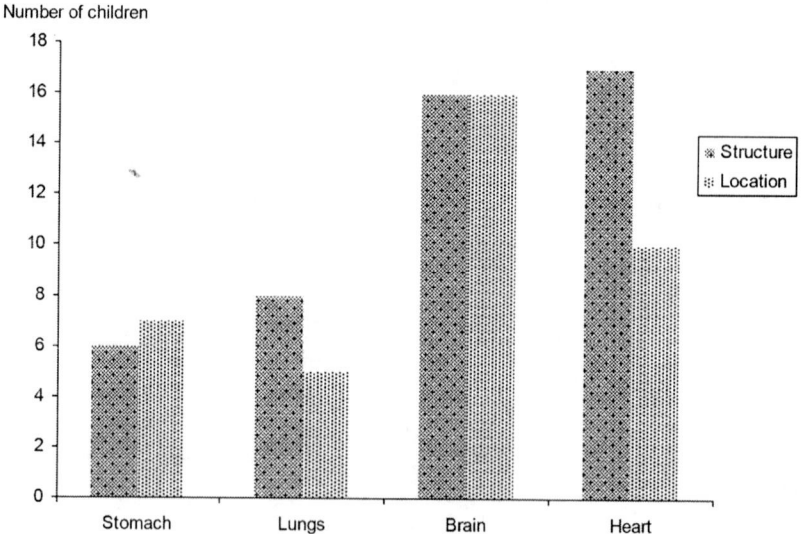

Figure 4.9 The number of children in P1 showing acceptable knowledge of the structure and location of four different organs as assessed by the drawings.

4.1.3 Heart and the Blood Circulation and the Lungs

Detailed teaching about the heart, lungs, and blood circulation did not take place until in Primary 2. It included talking about the heart being a muscle and that we could control some muscles but not others. The heart was an example of a muscle that we could not control. The children imitated the heartbeat by clapping the rhythm on their thigh. They did exercises where they had to run and jump up and down and count the heartbeat by putting the fingers on their own and on each other's pulse. They explored a model of the heart and listened to each other's heartbeat by using a real stethoscope. The children were asked to color a drawing of the heart by imitating the model of a real heart that was put on a table in front of them. When telling them about the blood circulation, the teacher put on a little act as if she was a blood cell traveling through the body and talked about the importance of the lungs in the process. She also read information from the textbook *Let's Look at the Body* (Óskarsdóttir & Hermannsdóttir, 2001a), where they looked at and talked about the picture of the heart and the blood circulation in the book. She talked about the difference between white and red blood cells, said that the red blood cells moved the oxygen around the body like a train, but the white blood cells were like guard dogs that fought bacteria and protected us against different diseases—phrases and words which are used in the text book.

In the drawings the children made afterwards of the heart and lungs and the blood circulation, most of them (15 children) drew a heart that was half blue and half red, just like the drawing in the textbook *Let's Look at the Body* (Óskarsdóttir & Hermannsdóttir, 2001a) (see Figure 4.10).

This time, however, just five children drew the heart like a "Valentine's" heart although the heart presented in the textbook is V-shaped. Most of the children started coloring blue veins on one side of the body on their drawing and red veins on the other side, but when the teacher saw that, she stopped them and told them again about the blood circulation, about the oxygen and how the red blood cells move the oxygen, and then the blood becomes darker because it lacks oxygen but gets more oxygen from the lungs. All the children then changed their drawings in the light of this information and drew red and blue veins all around the body. But it was not really until after this lesson that all the children knew that the blood traveled around the body in veins. The children were asked to draw the lungs on a transparency, cut them out, and place them in the right position on their drawing of the heart and the blood circulation. Most of the children (16 children) put the heart in relatively the right position, but nine put

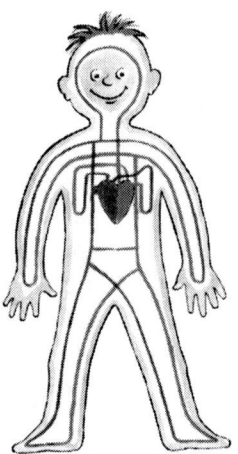

Figure 4.10 This drawing is adapted from the textbook used by the teacher. *Source: Let's Look at the Body,* Óskarsdóttir & Hermannsdóttir, 2001a, p. 14. Graphic by Sigrún Eldjárn.

the lungs either too low or too much to the right or left; some right on the shoulder (see Figure 4.11)!

The interviews showed that all the children identified the heart and knew its location in the model of the body (torso) that they had in front of them. All of them also knew that the heart pumped blood to the veins or to the rest of the body—that is, they knew the main function of the heart. One child mentioned that the blood kept us warm and three children

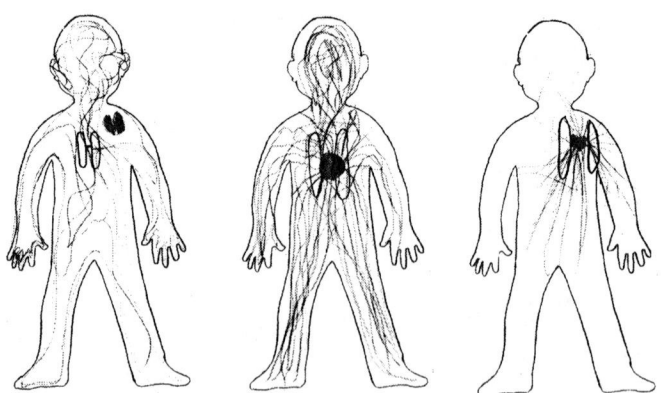

Figure 4.11 Three examples of children's drawings of the heart, lungs, and the blood circulation after teaching. Graphic by Sigrún Eldjárn.

mentioned oxygen in the blood. The lungs are connected to the ribs on the torso so it is not possible to look at the lungs separately. Most of the children knew that we use the lungs to breathe but they did not seem to know much more, or at least did not express more about the lungs. There were, however, two children who mentioned oxygen and that the blood got oxygen from the lungs. One child said the lungs cleaned the blood and one said that the lungs would get black if we smoked. Some also seemed to mix the white blood cells and carbon-dioxide-rich blood together.

The diagnostic tasks showed that all but two of the children knew the location of the heart. These two mixed the heart and the stomach together when they were asked to color the organs in varied colors. All of them, however, knew the function of the heart: "The heart pumps blood to the rest of the body." All but one of the children recognized the lungs and colored them blue as they were asked to do on a picture of the organs, and all but one (not the same one), "knew" that we use the lungs to get oxygen (in a false–true statement question), although most of them did not say anything about this in the interview when asked about the function of the lungs.

4.1.4 Digestion

Only two children drew the stomach on their initial drawing of the organs in the template provided at the very beginning of the project and another four children drew something that looks like the intestine although they did not know the name of it. When turning to the discussion about the digestion the teacher started by talking about the muscles that we cannot control. She told them that there were a lot of these muscles in the stomach and all the way from the mouth and through to the bottom. She talked about how important it was to chew the food well before swallowing it and about the teeth. She showed them the pictures on page 18 and 19 in *Let's Look at the Body* (Óskarsdóttir & Hermannsdóttir, 2001a) and discussed the way the food goes through the body. The picture on page 19 in the textbook shows a child sitting on the toilet. The teacher also told them that the nutrition from the food goes through the small intestine's "walls" and to the rest of the body. She asked the children why the muscles were working all the time and there were a few ideas, like: "The food digests," "The muscles are moving all the time." Then the teacher and the children sang the song. "All the food goes into the mouth and down to the stomach, so the stomach does not growl." After this the children were given a sheet of paper with a picture showing a child's face with an open mouth (Template 3.4) and they were asked to draw the food the child was going to eat and also draw the stomach and show how the food looks when it is there.

The teacher thought it was important to teach about the first half of the digestion first, talk about the food, the teeth, and how the food went down to the stomach and all about the little muscles working there although she also showed pictures and told the children about the rest of the process.

Only four children drew mixed or digested or half digested (in small pieces) food in the stomach. The rest of them drew the food whole, like whole carrots and whole slices of bread. When they had completed the drawings the teacher turned on the projector and showed them a picture on a big screen from the Internet (*Let's Look at the Body*, Óskarsdóttir & Hermannsdóttir, 2001b) that showed the way the food goes from mouth to anus. After that she did an interactive task with the whole class on the screen where one is supposed to pick an organ and put it in its right position. The children were surprised to see how small and how high up the stomach is located in the body. No one recognized the liver but someone thought it was the brain, by the structure of it. The teacher put the organs: heart, lungs, brain, stomach, small and large intestine, kidneys, and liver in their right places according to the children's suggestions. If the organ was put in the right place it remained there but if it was not put in the right place it did not stick and they had to try again.

When the teacher saw that so many children had drawn the food in the stomach in whole pieces she decided to talk to them again about this. One child was eating an apple and the teacher took the apple and asked the children if they thought that the apple went down to the stomach in a whole piece. "No, of course not," they said, and sounded offended: "Then you would choke and you could die." The teacher then asked them why they had drawn the food in whole pieces. One child said that they had drawn it like that because they did not know how to draw it otherwise, and he concluded, "We just did our best." The teacher then showed them again the pictures on page 18 and 19 in *Let's Look at the Body* (Óskarsdóttir & Hermannsdóttir, 2001a) and talked about the way the food went from the mouth and all the way through the body (see Figure 4.12).

A few months later (in Primary 2) the topic of digestion was brought up for discussion again. The teacher described in detail the route the food follows and used words like *a tooth paste tube* for describing how the muscles in the gullet push the food down to the stomach and how the muscles in the stomach work, like *a Kitchen Aid blender* that mixes the food into a mixture. She also talked about the way the food goes from the stomach and to the anus and talked about the organs involved in the digestion and how the nutrition goes through the intestine walls into the blood. Then she took the book, *The Magic School Bus* (Cole, 1996) and read for them from a chapter about the school bus traveling through the digestive system and what happens on the way. After

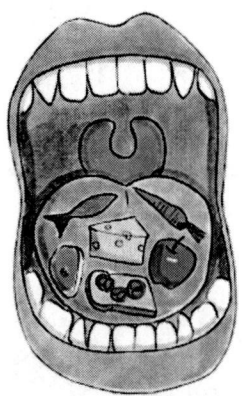

Figure 4.12 The drawing is adapted from the textbook used by the teacher. *Source: Let´s Look at the Body* (Óskarsdóttir & Hermannsdóttir, 2001a, p. 17). Graphic by Sigrún Eldjárn.

the reading she took two Weetabix cakes and put them in a transparent plastic bag and poured some milk into it. Then she walked between the children with the bag in her hands and her hands moving like muscles. She asked the children what they thought this was pretending to show and they all knew it was the stomach. The Weetabix and milk merged into a mixture very quickly. After this demonstration the teacher gave the children exactly the same kind of sheet (Template 3.4) as before (in Primary 1) and asked them to draw the food before it goes into the mouth and also how it looks in the stomach. She told them that they could draw the rest of the digestion system if they wanted, but they did not have to. All of the children drew the food as a mixture or very small pieces in the stomach this time and four children added more of the digestive system on their drawing (see Figure 4.13).

In the interview at the end many of the children (10 children) still had difficulties in recognizing the stomach in a torso because they thought it was much bigger than the model stomach in the torso showed. About half of them said the food was mixed or digested in the stomach but the rest, however, did not mention it. All but one of the children could show, more or less, the way the food went from mouth to anus, although they could not remember the names of the different organs involved; and two of them mentioned nutrition going from the mixed or digested food into the blood and to the rest of the body. The results from the diagnostic tasks show that all the children know that the food becomes a mixture in the stomach, although six of them are not sure about the way the food goes from mouth to anus and three of these thought the food went through the bladder and did not go through the colon at all. Three children mixed the stomach either

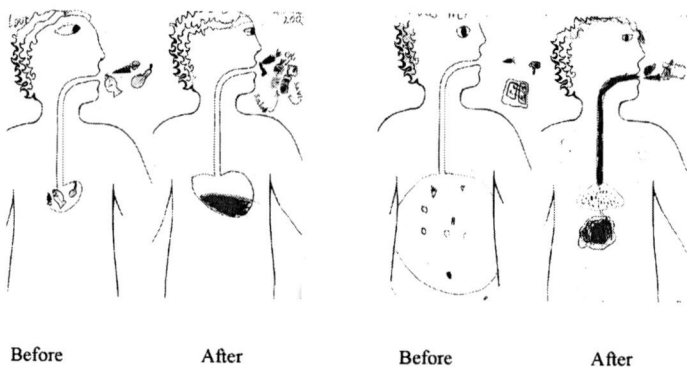

Before	After	Before	After

Figure 4.13 Examples of how two children demonstrate the food in the stomach before and after teaching about the stomach (Template 3.4).

with the liver or the heart, and five children did not know the difference between the colon and the intestine and mixed these two together, that is, both the structure and the location of the organs.

Consequently, the majority of the children did not draw the stomach or the digestive system on their initial drawing of the organs inside the body, and they thought the stomach was much bigger and located lower in the body than it really is. After intervention, many of them still had difficulties recognizing the stomach in a torso and three children were not sure about the stomach in a picture of the organs in the diagnostic task.

However, according to the interview, they did know more or less the location of the organs that make up the digestive system, or at least they could point at all the organs involved and describe the way the food goes from mouth to anus, although they could not mention all the names of the organs and were not sure about the structure of the stomach. But when they had got that right, the rest was easier and they all were aware of the function of the stomach and also the digestive process, though in a relatively simple way.

4.1.5 Brain

The results from the drawings show that, before teaching about the body started, twelve children drew the brain on their initial drawing (Template 3.2) where they were supposed to draw the organs inside the body. All of the nine children that scored at Level 5: Two internal organs (e.g., brain, heart, or stomach) in appropriate positions but no extensive relationships indicated between them, drew the brain and the heart as the two internal organs they knew (see Figure 4.6). A month later when asked to draw the organs again seventeen children (out of twenty) drew the brain. Between these two drawings there had been a class discussion in which the teacher was finding out the children's ideas about the body, where the children made a few comments about the brain, although in fact these came mainly from two boys, Óli and Árni, who shared their ideas with the rest of the class. Óli said that the brain controlled everything, and also said: "We could not live if we did not have a brain"; Árni said "Yes, and then we would not be able to think." Then Óli said again "We have a small control brain here below the main brain, it is connected to the main brain." Árni said the brain was not strong but the skull was like a helmet and if we fell on our head the skull would protect the brain; and Óli added, "Yes, the brain is both precious and sensitive/vulnerable." Then another boy joined in the discussion and said: "If you get a hole on your head the brain can crawl out." Here the teachers made a comment correcting the child, but said that it can be very dangerous to fall on the head and so it is important to wear a helmet when riding a bike.

The children did a few activities concerning their senses in the project as part of the discussion about the brain. These were about smelling different things, touching different materials, looking at different colors and tasting different types of food. They also did an activity focusing on their nerve cells where in a group they held each other's hands and made a "touch message" go through to the next person and so on. These activities also involved discussion on the connection between the senses and the brain. When teaching about the brain in more detail (in Primary 2) the teacher started by letting the children do a short memory exercise and then asked them what organ they were using to remember. The children were aware that they were using the brain. The teacher showed them the picture of the brain in the book *Let's Look at the Body* (Óskarsdóttir & Hermannsdóttir, 2001a) (see Figure 4.14) and read aloud for them information about the

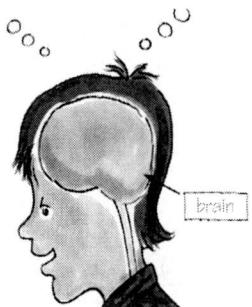

Figure 4.14 This drawing is adapted from the textbook used by the teacher.
Source: Let's Look at the Body, Óskarsdóttir & Hermannsdóttir, 2001a, p. 23). Graphic by Sigrún Eldjárn.

brain from the book. She talked about nerve cells and how messages travel from the nerve cells to the brain. Then Óli said: "There are really three brains, the main brain, the control brain that is here at the back, and the sleeping brain that is here right inside the brain. When I was in playschool I thought there was just one brain, but then I started thinking and saw a book with a picture of the brain and there you can see that the brain is like this."

The teacher had not planned to talk about the different parts of the brain but after this felt that she had to talk a little about the different parts of the brain. She got a book, *Svona erum við/That is How We Are* (Kaufman, 1976) and showed them a picture in the book showing the different parts of the brain colored in various colors. She told them that one part controlled eyesight, another part controlled hearing and so on. Then Óli concluded: "I know why we dream while we sleep and if we do not want to remember bad dreams we just have to wipe away the 'sleepy eyes' from our eyes."

After this lesson the children were asked to draw the brain. Results from their drawings show that two of the children colored the brain with three colors and many of them showed the brain divided into many parts (see Figure 4.15). Óli's brain was very detailed and with a "main brain" and a small "control brain" connected to it. Seven children drew the brain stem from one side of the brain, just like the picture of the brain in *Let's Look at the Body* (Óskarsdóttir & Hermannsdóttir, 2001a).

Results from the interview show that all the children (but one) knew the brain and its main function. One child (a girl, [10]) hardly said a word in the interview so it was hard to find out her ideas from the interview. The comments made by the children were: "We use the brain for thinking," "Messages are sent to the brain," "The brain helps us do maths," "The brain helps us concentrate," "The brain gets a message and it tells us to move,"

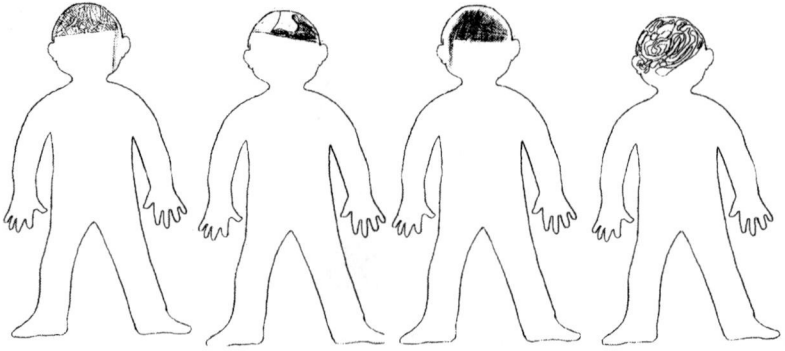

Figure 4.15 Four examples of drawings of the brain after teaching.

"The brain is divided into many different parts and there are strings from the brain and all around the body," "There is one part that is connected to the ear and the hearing, we think about something, it goes here and then we can say it," and "The brain helps you to know what you are doing."

Óli and Árni and another boy, Jack, were the only children that knew about the spinal cord although many children drew it on their brain drawing. Óli said that if the spinal cord got cut or hurt we would be paralyzed. Jack said the same and said the spine protected the spinal cord. Arni said that the spinal cord was a line and from the main line came many other lines that made you move: "I can just tell myself in my mind to move and then I just move" (and then he moved!).

Results from the diagnostic tasks show that all but two children know the main function of the brain (it controls the body, takes information from the senses, and helps you to be aware of danger) and that the skull protects the brain. According to the diagnostic tasks in which there are two questions/statements relating to this, only three children show that they know that nerve cells bring messages to the brain. Some of the children, however, confuse blood cells and nerve cells together. So the knowledge of the function of the brain in relation to the spine and the nerves and the processes involved, is vague.

As a result, after intervention and discussion where one boy in particular gave a lot of information on his ideas about the brain, many of the children had a broader knowledge about the function of the brain, and many of them drew the brain-stem as in the picture in the textbook *Let's Look at the Body* (Óskarsdóttir & Hermannsdóttir, 2001a).

4.1.6 Liver and Kidneys

In the class discussion at the beginning of the project when the teacher asked them initially what they knew about their bodies, no child mentioned the liver or the kidneys and no child drew the liver in their initial drawing of the organs. Two children drew the liver on the second drawing after the children had done the interactive task on the computer where, in pairs, they had to put the different organs in their right places. No one drew the kidneys on the initial drawing, but four children drew them on the second drawing. One of these four children drew the kidneys in the upper part of the body and another child drew them in the head. The child apparently remembered an organ that looked like this from the computer task but could not remember where it was supposed to be. (There is also a picture on page 20 in the text book, *Let's Look at the Body* [Óskarsdóttir & Hermannsdóttir, 2001a]).

In the interview, four of the children knew something about the liver: one said the liver is the body's main "cleaning factory," a phrase used in the textbook and by the teacher when talking about the liver, and another said that it can be difficult if the liver gets swollen or hurt. Three children (two of them the same children who knew about the liver) knew something about the kidneys also, although two of them did not know the word for it. One said the kidneys kept the blood clean, another said they had something to do with needing to urinate, and the third said they were glands.

There is no question or statement about the kidneys in the diagnostic tasks although they are included in a picture of the organs, but there is one task where the children are asked to color the different organs such as the liver. Fourteen children colored the liver correctly but the rest mixed the liver with the stomach.

Thus the results from the class discussion and drawings show that the children were initially not aware of the structure, location, or function of the liver and the kidneys. After a computer task, four of them drew the kidneys, and three of these drew the kidneys just somewhere in the body because they remembered that an organ that looked like this was in the body but did not know the name of it or what it was for, that is, the function of it. After teaching about the organs and mentioning the structure, location, and the function or purpose of the liver and the kidneys, the majority of the children still did not know about these organs.

4.1.7 Reproduction

The data on children's ideas about reproduction are derived from classroom observation (class discussion), interviews, and diagnostic tasks. No drawings were made relating to this issue. In the *National Curriculum Guide: Natural Science* (Menntamálaráðuneytið, 1999), reproduction is not a topic until in Primary 2. The guidelines say that children should know that babies are made by one cell from each parent. The teacher started the discussion about reproduction by asking the children how we were made, how our body was made. The children came up with different ideas like: "God created us," "The egg becomes a child," "We were animals in the beginning," "Once we were cells but then we became people," "Once we were almost apes," and another continued: "Yes, and one was the leader and they just used stones but then the brain gradually got bigger." Here they are obviously getting into the theory of evolution. The teacher asked the question again: "But how are the babies made?" Now there came more suggestions but now in a different direction: "Our mother gives birth to us," "There is a seed in the dad that comes into the mother and it changes into a child,"

"No, first it changes into a cell and then into an egg," "Yes, and then the child comes out and grows up."

After this discussion the teacher held up the big book, *Let's Look at the Body* (Óskarsdóttir & Hermannsdóttir, 2001a) showing them the pictures and reading aloud from the book how the sperm cell from the father and the egg cell from the mother become a baby. After looking at the picture in the book one child said: "The egg opens up and if two (sperm cells) can come in there will be twins"; but another said: "No, the cell comes from the egg," and yet another, "No, the cell is inside the egg until the cell turns into a baby."

The teacher again showed them the big picture from the textbook of an egg cell and sperms and told them how the sperms were trying to get into the egg and if one succeeded there would come a baby (see Figure 4.16).

She also told them about cell division and that the body was made of different types of cells that all came out of the egg cell and the sperm and these were: Blood cells, skin cells, muscle cells, bone cells, and nerve cells and they were all very, very important.

The data from classroom observation showed that children initially came up with mainly three different ideas about how babies are made, namely the idea that God created them, that human beings were monkeys in the past and gradually became men and women (evolution theory), and that babies were made by the father and the mother. After looking at the picture in *Lets Look at the Body* (Óskarsdóttir & Hermannsdóttir, 2001a), of a large egg cell and many small sperm cells that are heading for the big

Figure 4.16 This drawing is adapted from the textbook used by the teacher. *Source: Let's Look at the Body,* Óskarsdóttir & Hermannsdóttir, 2001a), p. 2). Graphic by Sigrún Eldjárn.

egg, and explanations from the teacher, the children tried to make sense of this information but had difficulties with the ideas of eggs and cells. It is as if they imagine the egg like a hen's egg inside the mother and find this difficult to fit into their ideas.

4.1.8 What Changed and What Did Not Change?

When looking at the development of children's ideas about the structure, location, and function of bones and the different organs of the body, many important results emerge from the data. The children had only a very vague idea about the bones in the body before the project started. But after the intervention, they all seemed to have a picture in their mind of the skeleton as a frame that held the body upright. They were aware of the structure and the function of the skeleton, even though they did not draw the bones connected together. They were, however, not aware of muscles being everywhere in the body but thought they were mainly on upper arms and legs and for doing physical things.

All the children knew the heart at the beginning of the project and the majority of them drew a V-shaped "Valentine's" heart. They also knew we have veins full of blood although some children imagined the body like a container full of blood. The second most frequently drawn organ was the brain. They also knew that the brain was in the head, and the heart somewhere in the chest. Thus they knew more or less about the location of these two organs. Their ideas of the function of the heart and the brain were very simple, as: "The heart pumps blood" and "It beats," and the children knew that we could not live without a heart. The also knew that we use the brain for thinking but very few seemed to connect the nerves to the brain.

After the intervention most of the children put the heart again in roughly the correct position, and this time only five drew the heart like a Valentine's heart, and all drew blue and red veins from the heart to the rest of the body. Now they all knew that the heart pumped blood to the veins and the majority knew the structure and location of the heart and all knew its main function.

The children's ideas about the lungs were very vague at the beginning of the project and only four children drew lungs on their initial drawing of the organs. After intervention, about half of the children still did not know the location of the lungs as some of them put the lungs either too low or too much to the right or left when asked to locate them in the body. They all knew, however, that we used the lungs to breathe although only a few of them had any knowledge of the blood bringing oxygen around the body. So the children were aware of the structure, location, and the main function of the heart (pumps blood to the rest of the body) and the main function of the lungs (breathing) although they were not sure about the structure and the exact location of the lungs and their knowledge and understanding of the processes that take place is very vague.

Only two children drew the stomach in the body in their initial drawing of the organs inside the body. After intervention many of the children (12 children out of 19) still had difficulties in recognizing the stomach in a torso because they thought it was much bigger than it really is. At the beginning of the project most of the children represented food in the stomach in whole pieces even though they knew they had to chew the food well before swallowing it. After intervention about digestion and the digestive system all the children knew that the food got into a mixture in the stomach and then went from the stomach through the digestive system and the rest out of the body through the bottom. After recognizing the stomach, the majority of them knew the structure and the location of the organs that make the digestive system, or at least they could point at the organs involved, although they could not state all the names of the organs. They were also aware of the main function of the stomach and also the digestive process, though most of them in a simple way. However, the digestive system and the digestive process were the most familiar compared to other organ systems and processes in the body.

The children were initially not aware of the structure, location, and function of the liver and the kidneys and after intervention the majority of the children still did not know about these organs.

The children had different ideas about how babies are made. They came up with answers and comments connected to religion, "God created us," but also to evolution theory, "We were animals in the beginning," and "Once we were almost apes." They also came up with the idea that babies were made by the father and the mother. After looking at the picture in *Let's Look at the Body* (Óskarsdóttir & Hermannsdóttir, 2001a) of a big egg cell and many small sperm cells that are heading for the big egg and explanations from the teacher, they were still quite confused and had difficulties when trying to make sense of these ideas about eggs and sperm cells. They seem to imagine the egg cell similar to a hen's egg inside the mother.

4.1.9 Information Obtained From Different Sources

When comparing the data obtained by the different methods, there are several things that have to be taken into consideration.

The drawings were collected to assess the development of children's ideas during the period in which the study took place. The children did the drawings as a natural part of the learning process but the data by interviews and diagnostic tasks were obtained at the end of the study. For this reason it is important to note that information obtained from drawings may not be fully comparable to the other methods because the children may have learned something between the time they did the drawings and the end of the project when the interviews and the tasks took place. This data from the drawings can however, give important information about the change in the children's ideas obtained at the early stages of the study and information obtained by the interviews and tasks at the end of the study. This must be kept in mind when the data are compared, but it gives a good opportunity to look at the changes that took place even though the data is not obtained by the same methods.

When information from the different sources is compared, it transpires that these three different ways to gather information about children's ideas, knowledge, and understanding may give different indications as to the children's knowledge. The results from the diagnostic tasks that the children did at the end of the study were significantly better in relation to all aspects of the stomach (location, structure, and function) than the results from the interviews ($p < 0.01$). This was also the case in the function of the lungs where the results from the diagnostic tasks were significantly better than the results form the interviews ($p < 0.01$).

The results give some important information about children's knowledge about structure, location, function, and processes and show that this varies according to the different organs. In this comparison, information is used from seventeen children that were both in Primary 1 and 2 and that data is available using all methods. This is put forward on Figures 4.17, Panels a, b, c, and d. Furthermore, as the digestive process is really the only process the children in the study were aware of, their ideas about processes are not included in the figures.

As can be seen on these figures, the three different sources give similar information about the children's ideas about the location and structure of the brain (Figure 4.17d). The children's ideas about the structure, location, and function of the heart are similar according to results from both the interviews and the diagnostic task. The tasks however show greater knowledge of the function of the lungs than the interviews. The greatest difference between information got from the three different methods is seen on Figure 4.17c, but according to the interviews the stomach was the least known organ as the children did not generally know its structure or location, even though all knew that we had a "tummy" the food we ate

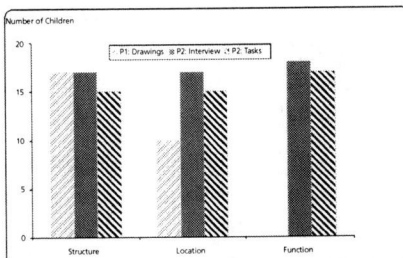

Panel a: Heart

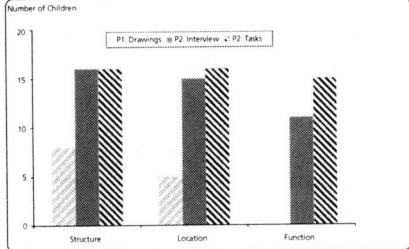

Panel b: Lungs

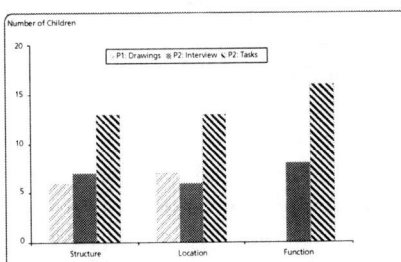

Panel c: Stomach

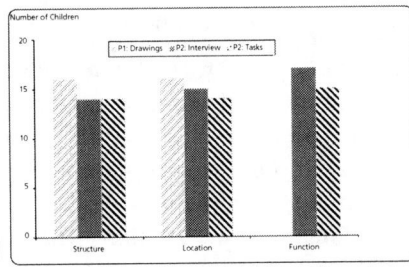

Panel d: Brain

Figure 4.17 Number of children showing acceptable knowledge of the structure, location, and function of the heart (Panel a), lungs (Panel b), stomach (Panel c), and brain (Panel d) as assessed by different methods. The drawings data were collected in Primary 1 (P1), whereas the interviews and the diagnostic tasks were used at the end of the study.

went into. As seen on Figure 4.17c, the "tasks" indicate greater knowledge of the stomach than the interview. The drawings show, however, there is little difference between them and the other methods regarding children's ideas about the structure of the heart and the brain, that is, their ideas did not change much according to this. On the other hand, there is considerable difference between the children's ideas about the structure and location of the lungs according to the drawings and the other methods, that is, the children have added to their knowledge of the lungs from the time the drawings were made. There is, however, very little difference between the results from the drawings of the stomach and the interviews, which suggests that in spite of detailed teaching about the stomach they still did not know the structure and the location of it in the interview. According to the results from the diagnostic tasks, which took place a week later than the interview, many of them knew the location and structure of the stomach, which suggests the children may have learned from the interview. This could also indicate that some information had been forgotten in the interview or the interview method was not suitable for extracting the relevant information. The first interpretation seems more reasonable, that is, the possibility that they learned from the interview, because of the high score on the diagnostic task.

Thus, as there is little difference between the results from the three methods about children's ideas about structure and location of the brain and the structure of the heart, it can be suggested that the children had the relevant ideas about these organs right from the early stages of the study. Their ideas about the lungs changed, however, from the time they did the drawings and to the end of the project, which suggests that the teaching about the lungs had been relevant.

When looking at the individual and when the results from the interviews and the diagnostic tasks are compared, it turns out that there are seven children whose results from the two methods are quite different. One child (10) scored full marks in the diagnostic tasks on the questions, statements, and pictures concerning the organs, but could not remember the names or the location of the organs and was very shy and confused in the interview. This child, however, had previously scored at Level 7 on the modified scale on the second drawing of the organs and at Level 6 on the second drawings of the skeleton on Reiss and Tunnicliffe's Scale (Reiss & Tunnicliffe, 1999a; Tunnicliffe & Reiss, 1999a) (Drawings 1–4, see table 3.1) Drawings are shown on Figure 4.18.

Another child also did very well in the diagnostic tasks but not at all well in the interview and said as little as he could (16). This child scored at Level 5 (skeleton) and Level 6 (organs) in the post drawings (Drawings 3 and 4, see Table 3.1). Maybe the same inference can be drawn here as in the previous case but here it is not as obvious because the scores are not as high. One child (20) scored quite low both in the interview and also on the diagnostic tasks. This child scored at Level 4 on the later drawing of the skeleton and Level 6 on the later drawing of the organs which indicates that he/she had not mastered the teaching introduced or had picked up elsewhere. Two children, however, (11 and 15) scored better in the interview than in the diagnostic tasks where their answers and marks were very confusing, and thus difficult to scale but one scored at Level 2 and the other at Level 4 in their later drawings of the skeleton, and one at Level 5 and the other at Level 6 in their later drawings of the organs. For them the interview seemed to be a more appropriate method.

There were two children that did especially well in the interview (5 and 13). One of these, Óli, did very well in all of the methods used to get access to the children's ideas. The other child (13) really bloomed in the

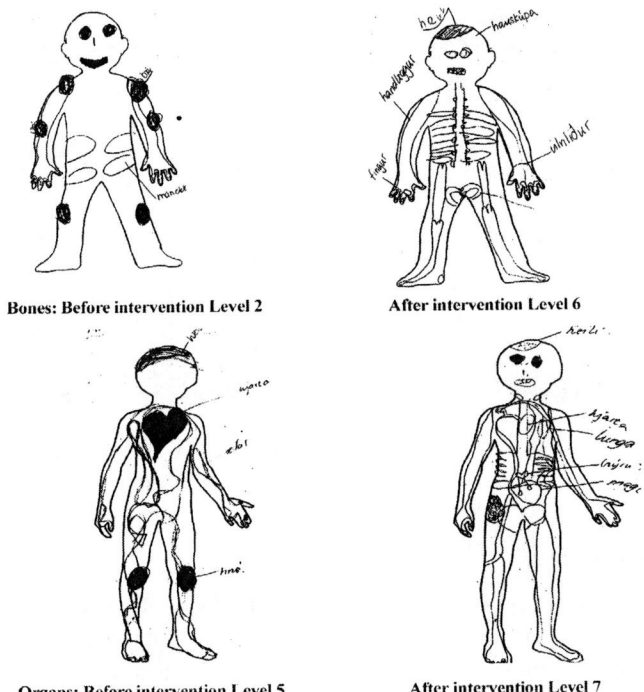

Bones: Before intervention Level 2   After intervention Level 6

Organs: Before intervention Level 5   After intervention Level 7

Figure 4.18 Drawings of the bones and organs made by Child 10 before and after intervention about the bones and organs (curricular Episode 1).

interview and gave much more information about his ideas, knowledge, and understanding than when the other methods were used. He scored at Level 4 on his later drawing of the skeleton and at Level 6 on his later drawing of the organs, but his drawings were difficult to analyze because they were so unclear (Figure 4.19).

This child got help reading the diagnostic tasks and needed much attention to keep him focused while doing the drawings and the diagnostic tasks. In the discussion and activities observed in the classroom he had difficulties concentrating and did not express his ideas a lot. However, in the interview he was very relaxed and his concentration was very good. He gave freely (more than I asked) a lot of information about his ideas, knowledge, and understanding and scored surprisingly high especially because he had not shown much interest in the project or made effort to express himself about issues involved in the project.

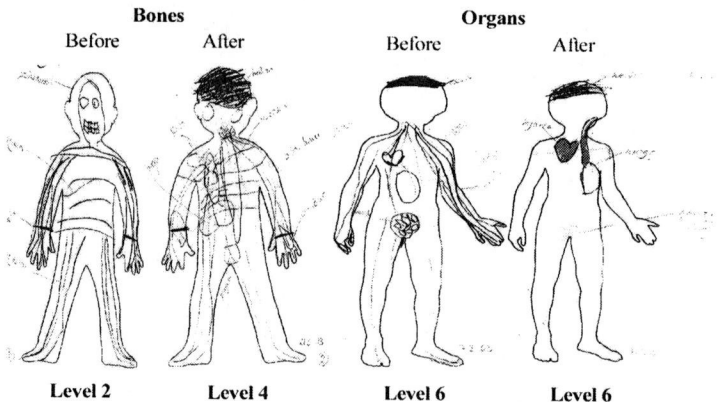

Figure 4.19 Drawings of the bones and organs made by Child 13 before and after intervention about the bones and organs (curricular Episode 1).

4.1.10 Óli's Ideas

One boy in the class has been mentioned by name several times before, Óli, who knows a lot about the body. When the teacher initially asked the children what they knew about the body, several times he said something about the brain. He said that the brain "controlled everything." He also said that we had a small brain below the main brain that he called "the control brain." He was the first child in the class who mentioned the word "organ" and knew what it meant. He also mentioned "pelvis" and pointed out where it was on a skeleton. In the drawings he made of the skeleton and the organs there is not much difference between the ones he drew before and after intervention. He scored at Level 6 in the former drawing of the skeleton but at Level 6/7 on the later, with only the hips missing. However, he scored at Level 6 in the former drawing of the organs, including a lot of organs (one among the few that drew the lungs and the stomach) but not connecting them together. In the later drawing he scored at Level 7 where he drew one organ system, namely, the digestive system. His drawings however, are not very clear, because his fine motor skills are not well developed compared to some of the other children, but he always wanted to explain his drawings to the teacher or to me, and so we wrote down the names of the organs and what he had to say about them. He was also the only child that said we grew "because of cell division."

In the interview when asked about how babies were made he said there were cells in "a sort of an egg container and the baby is in an egg inside the Mom." He also said: "The baby can not hear but it can understand when it is inside and when the Mom eats the baby gets food through the umbilical cord." The children were asked to draw a baby inside the mother's tummy. In his drawing there were a lot of pluses (+ + +) all around the baby. When asked about these pluses he said it was the oxygen and the child needed oxygen. In the interview he knew all the organs he was asked about, and also the liver and kidneys and their location. When asked about the function of the heart he said: "It is a muscle that pumps blood to the whole body," and he described the blood circulation quite correctly; and he connected heart and lungs, and described how we got oxygen into the blood from the lungs. He also said that we had to breathe and that we could not live if we did not have lungs. When asked to describe the digestion he explained how the food became a mixture in the stomach and then went from the stomach, to the small and large intestine, into the colon, and out through the bottom. He was very interested in the brain and said it was a "very important organ that controlled the whole body and if we did not have a brain we would just do silly things and we would not be able to live." He also said the brain was divided into three parts and pointed out three parts on the brain: "Look, this is the main brain, that really controls everything and sends messages to this one (points to the "small brain"), and that little one sends messages to the spine that sends it further to all the nerves and then you move" and he continued: "Look, the main brain has another part, here inside (pointed to a specific part in the middle of the brain), this is the "sleeping brain," it is so the main brain can rest."

Óli also said the brain was divided into many other parts that were very specific. When learning about the different cells earlier, Óli had said that we grew bigger because of "cell division." In the same lesson the teacher said to the children that dead skin cells fall off our body every day, and one boy in the class said that the dead skin cells went to the hair and turned into hair. The teacher said that this was not quite correct but the hair was really made of dead cells except for the roots. The discussion about this matter did not continue, but when interviewed, this was clearly something that had caught Óli's mind: It is as if he is trying to make it fit in with the ideas he has before because he says: "The things that you perhaps will forget maybe go out of your brain and through your head and through the hair like dead cells." Here he is trying to fit these ideas to his own but this is an example of a clear misconception.

When he was asked in the interview about other things he knew about the body he said the liver was the main "cleaning factory" of the body and the kidneys cleaned the blood. He also said that if the spinal cord got cut you would get paralyzed and concluded with, "The muscles make us move, we would be a pile of bones if we had no muscles."

Óli said that he had known most of this already when the project about the body started: "I knew it all before." When asked if he had not learned

anything new he admitted that he had learned some things better and especially the names of the organs. He said he had books about the body at
home and he had learned all this from books, science magazines, and TV
programs, and also from videos about the human body and all kinds of animals and about nature, and he concluded by saying that he wanted to be a
scientist when he grew up.

According to Óli's parents he had always been very interested in everything that had to do with nature and science. His parents are not scientists.
He has two older sisters and neither of them has been especially interested
in science or the body. His parent said that there was particularly one book
that he had explored when he was younger, *Svona erum við/That is How We
Are* (Kaufman, 1976). They also went regularly to the City Library where
he always took books about science and technology. He also borrowed
the magazine, *Living Sciences,* regularly from a friend's home. They also
said they had watched with him the film, *Osmosis Jones* (2001) and he had
greatly enjoyed it.

4.2　The Main Teaching Methods Used and Their Effects

There are many factors that potentially influence the changes in the ideas
of the children in the study. These include the effect the teacher has with
regard to the teaching methods and the teaching material she uses, the
teaching environment she creates, the tools and equipment she uses, and
the interaction that takes place in the classroom. In this section some issues
from previous Section 4.1 will be repeated but from a different perspective
as now the teaching methods are in focus. Attempts are made to keep the
overlap to a minimum but some are needed to make the sections coherent.

4.2.1　Teaching Methods

Data from classroom observation show that the teacher used a wide
range of teaching methods while teaching about the human body. Many
of these aimed at gathering evidence about children's ideas about the human body; others aimed at extending children's ideas. The teacher said her
main aim was to find out the ideas the children had about the body, the
bones, the different organs, and how it all worked, but also to "teach" them
about the body in order to add to and to develop their ideas. The main
teaching methods the teacher used were:

- Questioning strategies (question-and-answer sessions and class
discussion)

- Short introduction (mini lectures)
- Practical work/investigations
- Interactive activities on the Internet
- Drama
- Demonstration
- Drawings

The teacher used all these teaching methods but some more than others. She used short introductions (mini lectures) and question-and-answer sessions a lot. Usually she started the lesson with an open question, followed by a discussion where she tried to involve the children. Then she gave a short introduction to the topic in focus and then the children started working on different activities that aimed at helping them clarify and develop their ideas about the topic; and then the other teaching methods were used, that is, hands-on activities, demonstrations, and drama. She also asked the children to draw pictures of their ideas about different issues or topics before and after teaching about it.

When asked about the main methods she used she replied:

> That is really the main method I use. The rest is really in the hands of the children or at least they have influence and they give me new ideas to try with them.

When asked about the different teaching methods she used, she hesitated. It was as if she was not fully aware that she was using all these different methods and integrating different subjects. She was just teaching in the way she used to do,

> This is just the way I teach. I try to involve the children and try not to tell them everything myself and have it too teacher centred.

and she added:

> I am aware of the teaching methods I am using but I am not thinking about it all the time, I just try to use different methods to get all the children involved.

4.2.2 Short Introduction and Questioning Strategies—
Discussion Methods

Short introductions (mini lectures) and question-and-answer sessions were the most common teaching methods used. Short introductions could also be called "ignition" or "opener." It is something the teacher says and does to try to get the children interested and to catch their attention. When starting a lesson or opening up a discussion about an issue or a topic the teacher usually stood in front of the class holding the big book, *Let's Look at the Body* (Óskarsdóttir & Hermannsdóttir, 2001a). Sometimes she also used other books or showed them other things. For example, she used a big doll to get their attention when talking about the different parts of the body. She used a plastic model of a real heart when talking about the heart and she used a large paper model of a skeleton and also a big plastic model of a skeleton when talking and teaching them about the bones.

The teacher usually started the lesson or the topic by telling the children that they were going to look at something exciting. Then she usually asked a question to the whole class. She wanted the children to raise their hands if they wanted to say something or share their ideas and she asked the ones that did so. She usually repeated their answers to make sure that all the children heard and after a while there was a good sample of answers from the children. The teacher usually did not say if an answer was wrong but in order to clarify she repeated the answers and sometimes asked a new question to get a clearer answer. If the "right" answer did not come she usually gave them the answer herself or told them she would read about this from the book, which she then usually did.

If someone said the answer or gave the information she was looking for, she usually did not say so but continued encouraging the children to express themselves and share their ideas with the rest of the class. Sometimes she was looking for specific answers but usually she was just looking for the children's ideas about issues and concepts in order to gather evidence about their ideas. When the children that were willing to share their ideas had done so she usually added to the discussion, gave them extra information, and summed up what the children had said. She also often read aloud what was said about the issue in the big book about the body.

An Example of the First Class: A Discussion That Took Place Where the Teacher is Trying to Get Information From the Children About Their Ideas

> **Teacher:** Now we are going to learn about something exciting. We are going to learn about the body. What do we know about our body?
>
> **Lisa:** We have to go early to bed.
>
> **Óli:** The brain controls everything.
>
> **Tinna:** We have skin.
>
> **Árni:** There are insects or something inside us that digests the food.

Margaret: We have an appendix

Jón: We have bones, veins, and blood that floats.

Lisa: We have a skull.

Teacher: These are all very good ideas. Do you know something else?

Tinna: We have a heart so we can live.

Óli: If we did not have a brain we could not live. The brain is the main control part of the body.

Órni: Yes, then we would not be able to think.

Jón: We also have a face. If we did not have a face we would not be able to talk, see, and eat and we would not be able to find smell.

All of a sudden one child loses a tooth and this gets attention from the children that start talking about the teeth.

Óli: Teeth are stronger than bones.

Teacher: How do you know?

Óli: I saw it in a video of the Human body that I got from the City Library.

Tinna: The eyes are also very important.

Teacher: How?

Tinna: Because if we did not have eyes we would not see anything.

Teacher: Is there something else?

Lisa: The heart pumps.

Jón: The heart pumps fast.

Laura: If the heart pumps slowly we are dying.

Teacher: But what is it all called that is inside us. All the things that you have mentioned—the heart, lungs, brain, bones, lungs. Is there a common name for all of this?

Árni: It is just the body itself.

Óli: I know, it is called an organ.

Teacher: That is right, that is exactly the word I am looking for.

There are a few examples of direct corrections by the teacher. In one of the examples, when talking about how babies are made, one boy said that the sperms were insects. The teacher said, "No, it is called sperms, not insects." In another example, another boy said that cells were called "bells" (in Icelandic, frumur = þrumur). Then the teacher just corrected the words. This happened a few more times. The children used words that were quite like the words they were looking for but which meant something very different (in Icelandic, augngagn = gagnaugu, hnakkur = hnakki).

After one of the lessons when the teacher had used the question and discussion method a lot, she said she thought her teaching methods were too limited. She said:

> I think I use this method too much. I just give them direct information from this book (*Let's Look at the Body* [Óskarsdóttir & Hermannsdóttir, 2001a]) but that is really what I think a teacher should try to avoid.

She said by using this method she felt that she could get the children's attention and they seemed to listen and be interested in what she said and this supported her in using this method. She said she felt when using this method that she was in control and that made her feel good and "safe" and the teaching easy. She also said she had not always decided what questions to ask before the lessons started:

> It just happens, but I know what I am doing and I know the subject quite well. I am just trying to lead the discussion, get information about what the children know, sum it up, and give them more information. One thing just leads to another.

One example of the effect of the oral information given by the teacher in a lesson came when the teacher was explaining the function of the heart and the blood circulation. According to the drawings, most of the children did not understand the idea of oxygen traveling in the blood and how the red blood cells carry the oxygen and then the blood becomes darker because it lacks oxygen. At least they started to color blue veins on one side of the body on their drawing and red veins on the other side, and when the teacher saw that, she stopped them and explained and talked about it all again. All the children then changed their drawings in the light of this information and drew red and blue veins all around the body. And it was not really until after this lesson that all the children understood that the blood traveled around the body in veins even though their understanding of the circulation and the idea of oxygen in the blood was very vague. Another example where the class discussion along with existing knowledge had more effect than information from the teacher was when talking about the muscles in the body and the teacher told the children that there were muscles everywhere and we could control some of them and others not, and gave a few examples. This information about muscles being everywhere, did not seem to have much effect because some of the children also shared their ideas and said that we needed muscles to carry things, to move houses, and to win the strongest man of the world competition, and after this lesson most of them drew the muscles only on the upper arms and legs despite the teacher's information about muscles being everywhere. However, the teacher usually used other methods along with the discussion methods and books and pictures or something else in the same lesson, so that it is difficult to tell exactly if it was the discussion, the information from the teacher, or from the children or something else that had the most effect; but this will be discussed in more detail later.

4.2.3 Practical Work—Investigations

According to classroom observation, the teacher also used a lot of practical work when the children were engaged in different activities. For example, when the children were learning about the senses she had only half of the class at a time so that there were only two or three children working together on different hands-on activities, and all of them did all the activities. Here the teacher had the role of a leader or a helper, asking questions and telling the children to move on to the next activity when finished.

The children in both halves of the class seemed very interested in what they were doing and seemed to take an active part in the activities. The teacher said it was important to break the lessons up by doing something like this, getting the children engaged in hands-on activities or something different from sitting in their seats listening to her:

> These are the lessons that really make teaching worthwhile, when you see the children working happily and taking part in activities that interest them.

When learning about the heart, lungs, and the blood circulation, the teacher brought with her two real stethoscopes and a plastic model of the heart. She divided the children into pairs or groups of three and again they did different activities. They took turns in listening to each other's heartbeats through the stethoscope before and after jumping up and down. She helped them put the stethoscope on the right place on the chest and talked to them about their experience and findings. While some of the children were doing this activity under her supervision in the classroom the others were with me running up and down the stairs of the schools and jumping up and down in order to make them feel how the heart beats faster. The teacher said it was important to give the children opportunities to work together and also to try things on themselves, and "working with real stethoscopes like doctors do makes it all very interesting and exciting for the children."

According to the interviews and the diagnostic tasks, this activity with the stethoscope and listening to each others' heartbeat seems to have strengthened the children's ideas about the location and the main function of the heart, that is, pumping blood, although it is difficult to tell what exactly had the most effect here as the children knew beforehand that the heart ticks and keeps us alive as comments like this came from the children in the class discussion at the beginning of the project.

The teacher brought with her a model of a real heart, put it on a table where it could easily be seen by all the children, and asked them to color a detailed drawing of the heart using the model as a guide. According to the drawings, this activity, coloring a drawing of the model of the heart, did not seem to have a great effect. The children had the heart in front of them and were supposed to color the heart in the "right colors" but later when drawing the heart, lungs, and blood circulation, just four children drew the heart in any detail and most of them did not seem to have the model in mind when drawing the heart, because five children still drew the heart in a Valentine's shape and the great majority divided the heart into two halves and drew one part red and the other blue as in the school textbook. On the other hand, in the interviews when the children where asked to point out the heart after all the main organs had been taken out of the torso and put on a table, all the children identified the heart. So the coloring of the model of the heart might have had an effect after all, even though the children did not show that in their drawings.

When teaching about the bones and muscles, the teacher decided on a few different activities to widen the children's understanding. She made them work with and look at their own bodies by moving their joints and different parts of the body. They also walked up and down stairs in the school without bending their knees to see what it would be like not to have any joints. After the discussion about the bones and muscles she took them to the science classroom to look at and "explore" a full-size model of a skeleton and afterwards they completed, individually, a paper puzzle of the skeleton where they had to cut out the different parts and then put them together and glue them on a sheet of paper. She said this was important and one way of seeing what the children had learned.

The drawings the children made after this practical work show that it had a great effect on their ideas as all but two moved up two levels or more on Reiss and Tunnicliffe's scale (Reiss & Tunnicliffe, 1999a; Tunnicliffe & Reiss, 1999a), as seen on Figure 4.1. Ten children showed the joints on their post drawings but only one child (5) had drawn the joints on his initial drawing. However, the remaining ten children did not draw the joints in their post drawings. According to the drawings, the interviews, and the diagnostic tasks, all the children knew the location of the ribs and the skull, and that the ribs protect the heart and lungs, and the skull protects the brain. So the teaching that took place seems to have had an effect on these ideas.

When teaching about the senses the teacher had organized a workshop in the classroom with a variety of different activities and the children took turns in trying them all out. The teacher here had the role of an advisor or a helper, asking questions and telling the children to move on to the next activity when they had finished an activity. These activities, along with information from the teacher and information from the children during the discussion and interaction, seem to have extended the children's knowledge about the senses. This is borne out in four cases in the interviews when four children talked about the nerve cells and how touching something sends a message to the brain about how things are hot, sharp, or something else. However, the fact that the remaining 15 children did not mention the spine or the nerve cells in the interview or anything about how messages travel to the brain indicates that although they learned about their senses, few of them seem to have understood the relationship between the senses and the brain.

4.2.4 Interactive Activities on the Internet

While the children were doing a puzzle activity, putting together a skeleton, the teacher asked them, in pairs, to come and do a few interactive activities about the skeleton and the different organs on the two computers in the classroom. These were activities on the Internet, which are part of the teaching material used, *Let's Look at the Body* (Óskarsdóttir & Hermannsdóttir, 2001b). The children had to take turns in completing the activities. The teacher explained what they were supposed to do and then the children carried out the activities on their own or together as a pair. Some of them, however, had to get some extra help from the teacher. She said these kinds of activities made the teaching and learning more exciting and the children liked doing different activities and using the computer. She said the activities they were working on were fun to do but also an educational activity about the body. In one of the computer activities the children were supposed to put the different organs—the heart, lungs, stomach, kidneys, and liver—into the right place of the body.

According to the drawings the children made after this lesson, these activities seemed to have influenced their ideas, because in their second drawing of the organs inside the body, four of the children drew the kidneys just somewhere inside a drawing of a body, even though they could not remember the name of the organ or exactly where the kidneys were supposed to be. For example, one drew the kidneys in the head. This is an example of an issue that was not discussed or taught about especially.

4.2.5 Drama

The teacher used drama to enrich her teaching. When she was teaching about bones, muscles, and joints, she asked the children to imitate different animals—a giraffe, a seal, a dog, an elephant, and a gorilla. She controlled the play by leading them through the game and took part herself by bending herself back and forth. She also sang with them songs about the body that involve movement:

> Head, shoulder, knees and toes, knees and toes
> Eyes, ears, mouth, and nose,
> Head, shoulder, knees and toes, knees and toes.

It seemed easy for her to involve them and they knew all the songs she sang with them and took part in the singing. This, along with other activities about movement, seems to have had an effect on the children's ideas about the bones, muscles, and joints according to their drawings and diagnostic tasks even though it is difficult to tell exactly which had the most effect.

On one occasion the teacher divided the class by gender, and when teaching the girls about the different cells she used drama to help them understand cell division and also to understand how nerve cells sent messages to the brain and the brain back to the different parts of the body. The girls played the cells themselves and they seemed to really enjoy this and all seemed active and took part. The teacher was intending to use drama with the boys as well but they were so busy doing all the group activities exploring the senses that the lesson had finished before they were able to do the drama part.

The drama exercise about the nerves, however, did not seem to have any obvious effect to judge by the interviews, where I pointed at the spine on the torso and asked if they knew what it was while in the diagnostic tasks there were two questions/statements about the nerve cells bringing messages to the brain. Just four children knew about the role of the nerves according to both the interviews and the diagnostic tasks, and two of these were girls and two boys.

The teacher often used drama herself when she was explaining something to the children. Sometimes it was difficult to tell whether it was drama or demonstration or a mixture of both. When teaching them about blood circulation she took on a role herself where she changed her voice and made a little act like she was a little person traveling with the bloodstream, going into the lungs to get oxygen, and being pumped out from the heart again and again. By doing this she got the children's attention and she said to me subsequently "she could feel they were listening."

Results show from all the methods used to get access to children's ideas that, after the teaching about blood circulation, the children had quite a good idea about how the heart pumps the blood to the veins and how blood travels with the veins to the rest of the body, as discussed above. Whether it was the drama, the practical tasks, or something else is difficult to tell from the drawings, interviews, and the diagnostic tasks.

4.2.6 Demonstration

The teacher sometimes used demonstration to emphasize certain issues. A typical example is when she was teaching them about the digestion and put Weetabix and milk in a transparent plastic bag, walked between the children with her hands acting like the muscles in the stomach, and soon the Weetabix and the milk had mixed. The teacher had thought about bringing a food processor with her to school for the demonstration but afterwards she thought what she actually did was even better,

> because the children could see through the plastic bag and I told them my hands were working like the muscles in the stomach, this seemed to have a direct effect on the development of their ideas, they understood how the stomach works.

The teacher said that demonstrations like the one with the Weetabix and also all hands-on activities were very important for the children.

> Just little things like this is a change for the children, just like a good spice added to the recipe, they enjoy it and you just have to include this in the teaching.

This demonstration seemed to have an obvious effect on the children's drawings of digestion, for in the drawings they made afterwards all the children drew the food in the stomach as a mixture whereas in the drawings before, only four children had drawn the food mixed or in pieces in the stomach.

4.2.7 Drawings

The teacher asked the children to make a drawing before and after the discussion and teaching about each topic of the project (skeleton, organs, digestion . . .). The drawings gave important information about the development of the children's ideas. There were a few children who did not express themselves much in words so that it was impossible to tell just from the discussion what their ideas were and how they changed. The drawings give important information about that. The teacher said that: "Some children do not express themselves a lot so this is an important tool to see what they know." There was one foreign boy in particular who did not speak or understand much Icelandic but his drawings gave good information about his ideas. The teacher said: "Here you can see that he is learning from all the visual things like pictures, models, and the drama and that is a good feeling." Results from the drawings are discussed in other sections in more detail.

4.2.8 Teaching Material

The teaching material used was *Let's Look at the Body* (Oskarsdottir & Hermannsdottir, 2001a). The four teachers teaching the four Primary 1 classes in the school decided to use the material that had recently been published by the The National Centre for Educational Materials (NCEM) (Námsgagnastofnun). There is not much choice of Icelandic teaching material about the body. As *Let's Look at the Body* (Oskarsdottir & Hermannsdottir, 2001a) is the only material available, the teachers in Iceland really have no choice. The material is built on the aims and objectives in the *National Curriculum Guide: Natural Science* (Menntamálaráðuneytið, 1999).

The teaching material consists of two books. One is a large book used to display pictures to the whole class with additional text for the teacher. The other is a smaller textbook for the children with the same pictures but the text is very simple.

These materials are supplemented by an extensive website published by the NCEM (Námsgagnastofnun, www.nams.is) where a teacher can find teaching guidelines and other additional material, such as fuller information, different activities to try out with the class, a "Storyline frame" about the human body, interactive tasks on the computer for the children to work on individually or in pairs or small groups, games and drama exercises, and examples of formative assessment. Most of these can be printed out for use (see Óskarsdóttir and Hermannsdóttir, 2001b).

Teachers are encouraged to use this material as a source of information, and one page readily gives an idea for a topic that can last for several lessons. In the text there are words that obviously had an effect on children's ideas. The teacher used these words to describe different issues, as when talking about the liver she described it as really the body's cleaning machine. She also said the muscles in the stomach worked like a Kitchen-aid mixer, and the blood was a fluid that carried nutrition around the body like a train, with the white blood cells protecting us against illness like guard dogs—woof, woof . . .

According to the interviews, the children who knew something about the liver used these words and phrases, saying that the liver was some kind of a cleaning machine or had something to do with cleaning. One said: "The liver is the main cleaning factory of the body" and another said: "I can not remember the name of it but it cleans something." Many of the children mentioned in the interview that the food in the stomach got mixed because the stomach worked like a mixer: "The food mixes in the stomach, because the stomach is like a mixer." According to the diagnostic tasks, one child confused the white blood cells with "bacteria that fought like dogs" but also confused them with nerve cells where she wrote: "white blood cells" as an answer to the question: "What are the cells called that bring messages to the brain?" Another child also mixed the white blood cells with the nerve cells as the example below from one of the interviews shows.

Researcher: What is this (points at the brain in the model of the body)?

Child: The brain so you can think, it also controls.

Researcher: How does it control?

Child: I don´t know.

Researcher: Do you know what the cells are called that send messages to the brain?

Child: Yes, the white blood cells.

Researcher: No, they are called nerve cells, but what do the white blood cells do?

Child: They send bacteria away.

Researcher: Very good but what about the red blood cells?

Child: They come from the heart.

The drawings in the books have substantial effect on the children's ideas. The big egg (the mother cell) seems to confuse some of the children, because the egg in the picture is so like a hen's egg, even though it is shaped like a circle (see Figure 4.16). Some also think that the baby grows inside the egg or that the egg will change into a child as in this example: "The egg becomes a child."

The pictures explaining how the muscles work show the arm and muscles just in the upper arm which could be one of the reasons why the children drew the muscles just in the upper arms and the thighs (see Figure 4.20).

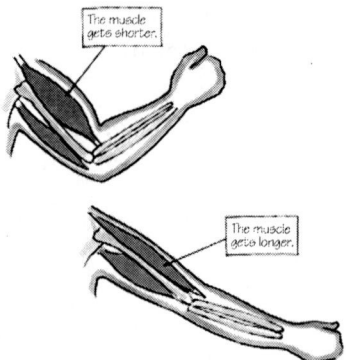

Figure 4.20 This drawing is adapted from the textbook used by the teacher. *Source: Let's Look at the Body*, Óskarsdóttir & Hermannsdóttir, 2001a, p. 10).

There is a large picture in the display book that is supposed to illustrate the blood circulation system (see Chapter 4.1). The picture shows the heart shaped like a Valentine's heart and red and blue veins around the body. However, where the red part of the heart is to the right, red veins are dominating, but blue veins dominate the left part of the body. This picture also seems to have considerable effect on many of the children. Even though the teacher has shown them a model of a real heart and talked about the heart, saying that it does not look like a Valentine's heart, in reality they still drew it like the heart in the picture in the book, half blue and half red; and five of them still drew it like a Valentine's heart. When asked to color the veins red and blue many of the children started by drawing blue veins in the left side of the body and red veins in the right side of the body as in the picture in the book. They changed it, however, when the teacher corrected them.

One picture in the book concerning the digestion system shows the mouth full of food in whole pieces, like a whole apple, or whole carrot, or a whole bread slice with cheese (see Chapter 4.1). When the children were asked to draw the food in their stomach before the teacher really taught them about the digestion system, most of them drew the food in their stomach in whole pieces as in the picture, even though they knew, according to the discussion, that you would choke if you swallowed the food in whole pieces.

There are also very detailed pictures of the kidneys in the book, which could, along with the interactive tasks on the Internet, have had some influence on the children who drew kidneys in their drawings of the organs in the body. There are also two similar pictures of the brain in the book:

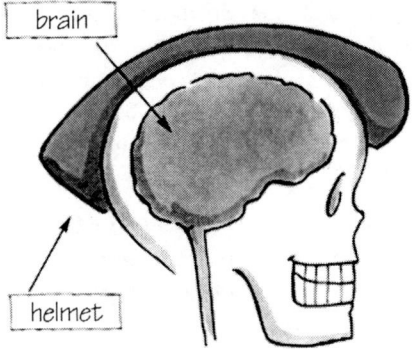

brain

helmet

Figure 4.21 The drawing is adapted from the textbook used by the teacher. *Source: Let's Look at the Body*, Óskarsdóttir & Hermannsdóttir, 2001a, p. 9). Graphic by Sigrún Eldjárn.

both show a gray brain with a brain stem down one side of the brain (see Figure 4.21). When the children drew the brain many of them drew it like the brain in the book with the stem down one side.

The teacher also used the books, *The Magic School Bus: Inside the Human Body* (Cole, 1996) and *Svona erum við/ That is How We Are* (Kaufman, 1976) to emphasize her points and add to the discussion. She used *The Magic School Bus: Inside the Human Body* (Cole, 1996) when explaining the blood circulation and the way the food goes through the body when she put on little drama acts. She used *That is How We Are* (Kaufman, 1976), when talking about the brain and showed them a detailed picture of the brain in the book and some of the children had that picture clearly in mind when drawing their own picture of the brain.

4.2.9 Summary

The teacher used a wide range of teaching methods: short introduction (mini lectures) and questioning strategies (questioning and discussing), practical work/investigations, interactive activities on the Internet, drama, demonstrations, and drawings. All of these aimed at gathering information about children's ideas about the body, and some also aimed at adding extra information and experience and extending their ideas. The teacher knows the subject and she is familiar with and uses the different teaching methods with confidence. She is not used to classifying and talking about the methods she is using, but she knows them well and employs them without thinking about what methods she is using. "This is just the way I teach," she said. She is aware of the importance of using different teaching methods although she uses questions, discussion, and short introductions most, and is rather worried about the large role these methods play in her teaching. She uses these teaching methods because then she feels secure and in charge of the whole situation. However, she also uses other methods but more to break up her "traditional" methods and make the learning more interesting and fun for the children.

The different teaching methods seem to have had an effect on the development of the children's ideas although it is difficult to tell exactly which has had the most effect because the teaching methods were not used in isolation, but in every lesson and for every topic the teacher used several. There are, however, examples that clearly had an effect on their development of their ideas and even changed them; one is the demonstration she did when teaching about the digestive system, and another was the interactive tasks on the Internet where they had to put the organs into the right places. There were also clear differences in their drawings after the practical activities they did with their bones, muscles and joints, bending, and imitating different animals. Also, after the discussion about the bones and muscles where the children believed the muscles were important if we have to do something physical like moving or carrying heavy things, most of them drew the muscles only on the upper arms and legs even though they were told that there were muscles all around the body. This is an example where the teaching did not have the intended effect.

The activities the children did in connection with their heart, jumping up and down and listening to their heartbeat through the stethoscope, seemed at least to clarify their ideas although it is difficult to tell which exactly had the most effect. However, it seems from the evidence here, that discussion along with demonstration with different visual aids and practical activities that the children take part in, are more effective than discussion alone.

The teaching material *Let's Look at the Body* (Óskarsdóttir, & Hermannsdóttir, 2001a) clearly also had a great effect on the children's drawings because they tended to imitate the pictures in the book when they were asked to show their ideas in drawings, although the drawings did not always represent their ideas as revealed in the interviews and diagnostic tasks.

4.3 Pupil Involvement and the Interaction in the Classroom

In this section the focus is on involvement and interaction in the classroom. The results are based on data mainly from classroom observation but also from interviews. The teacher used different methods to get her pupils

TABLE 4.1 How Children Were Mapped on the Quietness Scale

	Quietness Scale							
	1	**2**	**3**	**4**	**5**	**6**	**7**	**8**
Child Number	5	17	1	13	7	6	11	9
	15	22		19	8	21	14	10
					18			12
								16
								20

Note: 1 = Most Active; 8 = Most Quiet

"active" or engaged in the learning process. I realized early in the process of my study that "being active" (In Icelandic–að vera virkur, að taka virkan þátt) was an important concept in my research.

As discussed in Section 3.6, the children were divided into 8 levels on the quietness scale (see Table 4.1). Child 21 and 22 were not in Primary 1 and are therefore not included in the data on many of the drawings.

When the children were divided into three groups the distribution was as follows:

- Level 1–3 = The *visibly active* group (4 boys and 1 girl, the girl on Level 3)
- Level 4–6 = The *semi-active* group (3 boys and 4 girls)
- Level 7–8 = The *visibly passive* group (3 boys and 4 girls)

According to the criteria above, it could be inaccurate to talk about the class discussion. However, the teacher tried to involve more than just the children in the active group and the children in the semi-active group sometimes took part too.

Both the teacher and I seemed to use similar criteria when deciding if a child was paying attention or not. However, it can be difficult to determine this. The teacher sometimes said that "she could feel they were listening" and I, the observer, being in the classroom also "could feel and see" whether they were paying attention or not. This may be vague but this was the criterion we used. The teacher tried to involve the children who did not raise their hands in order to share ideas and were not taking part in the discussion. Sometimes she just asked a particular child directly or maybe two children to get them involved or just to encourage them to add something to the discussion. Usually they did not respond but occasionally they did. She said it was important to involve them even though they did not say a lot:

At least I feel I have tried to involve them and I believe that even though they do not say a lot, they listen to me and the others and are hopefully learning something.

4.3.1 *Difference Between the Involvement of Boys and Girls*

I had written in my notes a few times that the boys were much more active than the girls, that is, they expressed themselves more and were more willing to talk and share their ideas. In the very first lesson when the teacher asked the children what they knew about the body I soon learned a few names. These were the names of the children who generally wanted to take part and share their ideas. In this first lesson the boys and girls were equally active in the discussion. In some of the other lessons, however, the boys were much more dominant, and sometimes there was really a sharing of ideas and discussion between only two or three. The teacher however always tried to involve the rest, and some occasionally came in with one or two comments. This was the case in the discussion about reproduction (see Chapter 4.1.7) and also in the lesson about muscles that we cannot control, as in this example:

Óli: Yes, when we swallow.

Teacher: Yes, that is correct, when we swallow. Some other ideas?

Árni: Yes, there is a certain body in the head that makes us do lots of things (he moves himself).

Teacher: A body?

Árni: No, it is the brain.

Óli: There is a muscle in your throat to help you swallow.

Jón: Yes, I know, it is called Adam's apple.

Teacher: Yes, it is sometimes called the Adam's apple (some of the children have obviously heard this before because they agree, but it is not commonly used in Icelandic). But do you know what is the body's biggest muscle?

Óli: The calf on the leg.

Teacher: No, but that is a big muscle.

Árni: The thigh.

Teacher: That is also a big muscle but the biggest muscle is on the bottom.

The children thought this was funny.

Teacher: We have muscles inside our body that are always moving but
we do not notice that. There is, for example, one that this will
remind you of. (She claps with her hand on her thigh few times).

Margaret: I know, the heart.

When asked about the difference in engagement between boys and girls the teacher said she thought it was about "fifty/fifty" (six children—three boys and three girls) that were more engaged or active in the discussion than other children. I can agree with her up to a point when she says that there are only six or even five children really involved in the discussion, but on the basis of my notes I can see that the boys were generally more willing to take part than the girls. According to my notes from the classroom observation it is really one girl and four boys that take an active part in the discussion and are in the visibly active group, but there are another three girls that sometimes take part and are therefore in the semi-active group. Especially the four boys seemed more willing to express their ideas than the rest of the children and seemed to be more interested in the topic. The girls, however, were generally quieter, hardly ever expressing their ideas without raising their hands, and were more passive, meaning that they were not just quiet but also did not seem to be so interested or enthusiastic about the project, or at least did not show it in the same way as some of the boys did (i.e., raising their hands, calling the answer out or saying: "I know, I know" and wanting to answer or share their ideas). This can often be seen in my notes and transcripts of the discussions.

The teacher divided the class once into two groups by gender. This was when the children were working on practical activities about the senses. When the girls were alone all of them were much more active than when the whole group was together. Those girls, who had not seemed so interested before, now seemed very interested and took part in all the activities—drama and hands on activities where they were exploring their senses. They worked in small groups of 3, 3, and 4 and all the groups had to be led through the activities by the teacher. None had the initiative to start working on the activities on their own, even though they seemed interested and content. They merely followed the instruction from the teacher and did what they were told to do but did not express themselves much during their work. However, without the boys, all the girls including the three girls in the semi-active group were more active now than in the other lessons when all the class was together. Then, some seemed to be in a world of their own—drawing or coloring. It is difficult to tell just from the observation whether they were listening or following what was happening, and most were reluctant to say what they thought or to express their ideas.

This did not apply only to the girls. There are also a few boys (3–4) that hardly ever added anything to a discussion, and two of them seemed very shy. In the lesson where they were alone without the girls, they were generally more active than usual. The boys were generally more relaxed in this lesson when they were in all-boys groups than the other lessons when the whole class was together. Although they all took an active part in the activities in all-boys groups, the more passive boys followed the others, and the same three or four boys who normally dominated the discussion still did.

The boys were generally very enthusiastic and eager to try all the different things and seemed as a group to be more independent of the teacher than the girls. But they also argued, and some of them on one occasion spilt on the floor the material they were working with and had to get a brush to clean it up. Two boys started fighting over one activity and argued about who should be the first to try, so there were also some discipline issues involved.

In the notes I wrote after these lessons when the boys and girls were segregated, I said it had surprised me how little difference there was between the behavior of the boys and the girls because although there was more action among the boys and they were louder, both groups seemed to get a lot out of the lesson and all the children were more active here than in lessons when the whole class was together. I had also written, that because the teacher was organizing the same lesson again, repeating it for the boys, at the beginning they were given more detailed instructions on what to do and the content of the different tasks than the girls, and that seemed to have helped the boys and made them more confident. Thus, both boys and girls seemed to take generally more part when divided into groups by gender than when the whole group was together.

4.3.2 Being Active Means a Lot More Than Taking Part in the Classroom Discussion or Expressing Ideas

According to the drawings and to the diagnostic tasks, being active in the sense of learning can mean a lot more than taking part in the discussion or expressing ideas. By looking at the drawings, having in mind the teaching methods, the visual aids (for example pictures and models), and the discussion that took place, it is clear that the children's drawings have changed a lot from the beginning of the project to the end. Even though it is not clear what may be attributed to imitation of the pictures in the textbook used, one can venture to infer that their ideas have changed substantially. However, while the changes may be due to activities at school (in class), they can also be due to maturation and to the teaching they have obtained at home or elsewhere.

According to the results, a child can be active, in the sense of learning, by looking, by listening, by trying out things, and by completing activities. Some of the children who never said anything to the whole class, who were shy or just did not want to share their ideas, that is, the children in the visibly passive group, made drawings that show how their ideas changed from the beginning of the project to the end, and so it can be assumed that they have been active in some way or another and have been learning something.

In the lesson when the focus was on the heart, lungs, and blood circulation, two children (a boy and a girl) were not listening at all. They were giggling and poking into each other. Two other children (also a boy and a girl) sitting together were at the same time not taking any visible part in the lesson either. They were quiet and seemed as if they were thinking about something else. According to the drawings that these two quiet children made afterwards compared with the ones they did earlier, they seem to have been listening and watching after all, and therefore they had been active, in the sense of learning, even though they were not taking part in the discussion. In their previous pictures, the boy had drawn the lungs very low in the body but the girl had not drawn any lungs at all and her heart was Valentine shaped. The hearts in their drawings now were much more detailed (especially the boy's heart) and the lungs were in the right place.

The drawings made by the two giggling children showed that they were not paying attention and not taking in what was being presented and discussed. At least there were no signs of learning in the drawings they made at the end of the lesson. When asked to comment on this the teacher said:

> It is difficult to say whether some of them are learning or not because they do not take part in the lesson and are so passive. They are just doing their own things and it is difficult to tell when someone is learning. Some children just learn by listening, they have no interest or need to say anything; they just listen and do their work.

She also said that it was the same story with all the different teaching methods: "no one method fits all." She also said it was difficult to say whether the 60–70% of the children in the class that were not taking part in the discussion were getting anything out of the discussion:

> It is difficult to tell. Maybe it is better for them when I stand in front of them and read from the book or tell them about all the different things. Just like some children that just love to listen to stories and you can teach a

lot through stories and there are a lot of interesting books that you can use in your teaching. Maybe this is the best way for some children. There is no correct or best way of teaching all children.

When trying to get her to clarify her ideas about children's engagement and the proportion of children that are really involved and taking part in the classroom discussion, the teacher said again that there were usually just around 30% of the children that are visibly active (5–6 children) and taking part. She also said these were usually the same 5–6 children and that this was also the case in other subjects she taught them and in other situations, always the same children engaged or active. She said that according to other teachers that taught the class this was also the case.

4.3.3 Are the Quiet Children *(the Visibly Passive Group)* *Learning Less Than the Others?*

When looking at the data obtained from different sources: drawings, interviews, and diagnostic tasks, and at the three groups of children: the active group (5), the semi-active group (7), and the visibly passive group—the quiet children (7), some important issues emerge.

Note that I only include the 19 children that were in Primary 2. Although the visibly passive group (the quiet children), have in common being quiet, their drawings, interviews, and diagnostic tasks show very different results. There were three quiet children that showed substantial difference in ideas between the beginning and end of the project. One of them, a foreign boy that never spoke a word in class discussions and understood very little Icelandic when he started school, did change from Level 2 to Level 4 on the skeleton scale and from Level 5 to Level 7 on the organ scale. He did very well in the diagnostic tasks where he was the only child who put the correct tag to each bone on the skeleton, and he also did quite well in the interview.

There was also a great difference between the initial drawings and the second drawings of the other three quiet children. One of them (Child 10, see drawings in Chapter 4.1.9, Figure 4.18) changed from Level 2 to Level 6 in the drawing of the skeleton and from Level 5 to Level 7 in the drawing of the organs. This child (a girl) did very well in her drawing tasks and also in the diagnostic tasks but not at all well in the interview, where she was very confused and did not know either the structure or location of organs. The other two children's drawings (Children 14 and 16) changed from Level 2 to Level 5 in the drawing of the skeleton and Child 14, a girl, changed from Level 3 to Level 6 and Child 16, a boy, from Level 5 to Level 6 in drawings of the organs. The boy did very well on the diagnostic tasks as he knew the structure and function of the organs and the digestion process, but like the girl, did not show what he knew in the interview. He was very shy and just did not answer. Thus, the interview did not provide the same information about the quiet children's knowledge as the other two methods used (drawings and diagnostic tasks).

There was one other quiet child, a girl (12), whose position was similar to these two in that she did well in her drawings and diagnostic tasks but did not say much in the interview; but as she had been absent in some lessons I did not have all her drawings. The last two quiet children (11 and 20) were a boy and a girl. The girl did not seem to have learned a lot according to the interviews and diagnostic tasks, or at least it was difficult to see this from the data. Her organs drawings showed, however, that she moved from Level 3 to Level 6 although she did not seem to know the structure or the function of the organs in the interview. The boy was very shy and did not say more than single words and short sentences in the interview. However he pointed out the correct organs when asked to and put them in the right places, but his performance on the diagnostic tasks was especially poor as he did not answer or react to many of the questions and did not complete the activities, although he could answer the questions about the function of the different organs. Accordingly, it is impossible to say that the quiet children have anything more in common than the fact of being quiet. In Figure 4.22, it can at least been seen that from their drawings of the skeleton and the organs inside the body, the intervention that took place between the two sets of drawings has had considerable effect: It means that they were learning even though they did not take any part in the discussion or express their ideas orally.

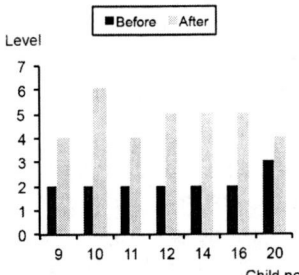 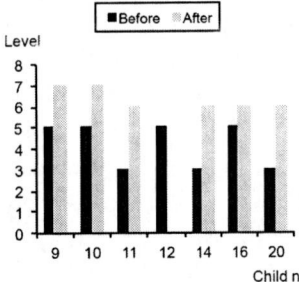

Figure 4.22 The quiet children. Drawings of the bones (left) and organs (right) made before and after teaching.

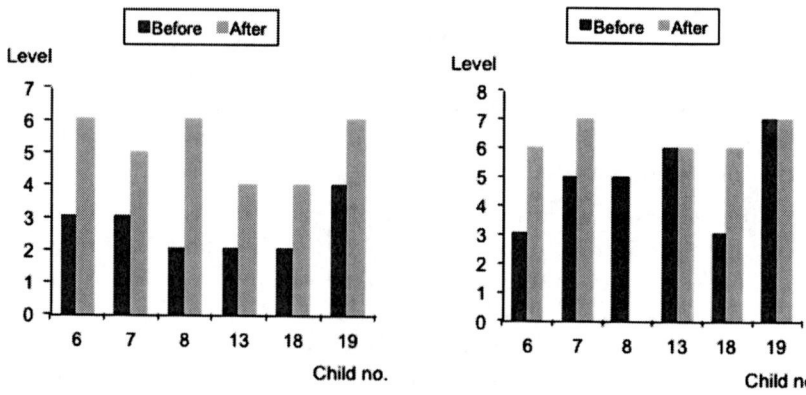

Figure 4.23 The semi-active children. Drawings of the bones (left) and organs (right) made before and after teaching.

The semi-active group contains seven children. Figure 4.23 shows only six children as the seventh child was not in Primary 1 and is therefore not included in the data on many of the drawings and is therefore not included in Figure 4.23. The drawings, interviews, and diagnostic tasks gave similar information about the ideas of these children. The children in this group knew the structure and location of most of the organs and also their function. There was one child, however, in the semi-active group who had difficulty with concentration. It was difficult to analyze his drawings because they were messy. Even though he was given help reading through the diagnostic tasks, the results together with the drawings did not show that he had learned much during the project, although he knew about the four main organs: heart, lungs, stomach, and brain (see Figure 4.19). However, he blossomed in the interview, where he really showed a lot of knowledge about the function of the different organs and a lot of extra relevant matter too. So here there was a great difference between the results from the methods used to get access to his ideas. But with this exception, the results generally of the children within the semi-active group were quite similar, and the methods used to get access to their ideas gave similar results.

There were five children in the visibly active group, four boys and one girl. One of the children classified in the group was not in Primary 1 and is therefore not included in the data on many of the drawings and is therefore not included in Figure 4.24. The four remaining children did not have a lot in common. One of them, a boy (15), really liked to share his ideas and he knew quite a lot about different things concerning the body, though usually not ideas that the teacher was working with. However, he often came up

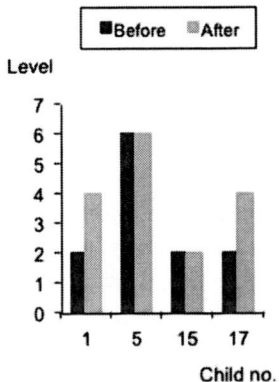 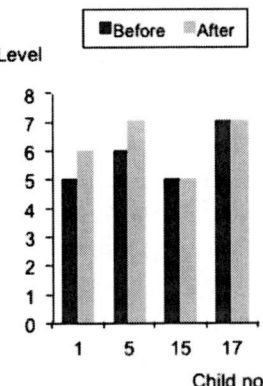

Figure 4.24 The visibly active children. Drawings of the bones (left) and organs (right) made by the active children before and after teaching.

with good ideas and information even when not quite relevant. He got help while going through the diagnostic tasks but did not understand some of the tasks. His drawings, the interview, and the diagnostic tasks showed that he had only vague ideas about the structure and functions of the different organs, and he was confused in all his ideas about the body. These two boys often wanted to express their ideas but they were not always relevant to the issues in focus.

The three remaining children are Óli, Árni, and Margaret (5, 17, and 1). Margaret was the first girl's name I learned. She seemed very open and cheerful and willing to take part in the discussion and share her ideas, but in a polite way. She always raised her hand and waited for the teacher to give her a sign to talk during the class discussion, but compared with the visibly active boys she did not express herself as often as they did. In the class discussion she seemed to have a lot to contribute to the conversation. In light of that experience, I thought she would show a better knowledge in the interview than she did. Her drawings on the skeleton moved from Level 2 to Level 4, but her drawings on the organs just moved up one level from Level 5 to Level 6. According to the interview and the diagnostic tasks, she knew the structure, location, and function of most of the organs but she did not express herself much in the interview, just answering the questions in single words and short sentences, even when I asked her open questions to try to get her to talk about her ideas.

Árni is in a way quite similar to Margaret in the sense that in the class discussion he had a lot to contribute and seemed very interested in participating in the discussion, and their scorings in their drawings of the skeleton were the same both before and after the intervention, both scoring at Level 2 before and at Level 4 after intervention. However, Árni scored at Level 7 on both drawings of the organs. In the interview and the diagnostic tasks he did however, less well than I had expected compared with the ideas he shared in the classroom discussion; he knew the structure and function of most of the organs but his ideas about the digestive process were more vague, as neither in the interview nor in the diagnostic task was he able to show how the food went from mouth to anus. So the visible active children are not necessarily learning more than the less visibly active ones.

Óli's ideas were described in Chapter 4.1 and his influence on the other children´s ideas will be discussed later in this section.

In summary, the results indicate that the ideas of the children within the visibly active group are different from one another, as in the case of the children within the visibly passive group. The ideas held by the children in the semi-active group are more similar to one another, and the various methods used, drawings, interviews, and diagnostic tasks, give similar information about their ideas.

When the pre- and post-drawings of the children in the three groups are compared, it can be seen that the visibly passive and the semi-active groups perform similarly. Similar progression is to be found in both groups. But the visibly active group changes differently from the rest, that is, the interaction between time (before and after) and activity level is significant, $F(2,12) = 7.1$, $p < 0.01$. The active group shows a different pattern than the rest. Their gain is not as great as that shown by the others (see Figure 4.25). The high scoring at the beginning may suggest that the lower gain may be due to a ceiling effect, indicating that perhaps they knew at the beginning of the project so much that they had little to gain according to the ceiling set by scale. The data for the bones can be used to reject that explanation. Another possible explanation could be that the teacher did not present the children with the opportunity to learn more. Similarly the bones data undermined this explanation. The third possible explanation suggests that their activity actually hindered their learning; an explanation that the data does not address.

There is a problem in interpreting the data that concerns the active group. There are only four pupils who are so classified and even though the critical interaction is statistically significant it is somewhat tenuous to base crucial interpretation on so few individuals. However, it is clear that the results show that the active ones are not learning more than the others and the quiet children are not learning less than others and in the data there are strong indications that the opposite, that is, the passive children

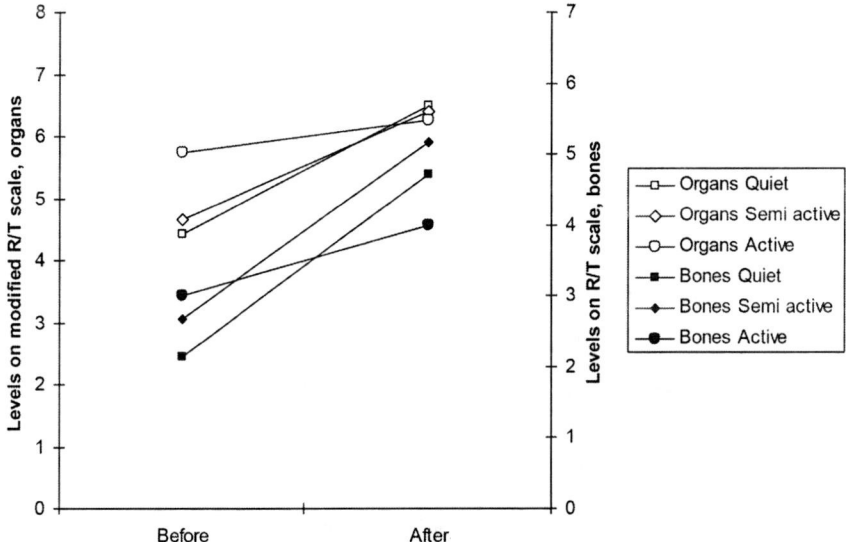

Figure 4.25 The levels attained by the pupils before and after teaching about bones and organs based on drawings in Primary 1 from the three activity groups. *Note:* The scale for organs extends from 1–8 but the scale for bone extends from 1–7.

are learning more than the active children. Another line of argument is to use a slightly more refined classification of activity. When this is done the picture is confirmed and somewhat clarified. Figure 4.26 shows the scattergram for the gains on the bones and organ scale averaged. The quietness scale can explain about half ($R^2 = 0.50$, $p < 0.02$) of the gains scores, which suggests that that quietness is correlated with greater gain is supported, though of course this needs further substantiation.

4.3.4 The Influence the Children Have on Each Other's Ideas

There are some instances that indicate that sharing of ideas and what children say in the class discussion could have had an effect on other children's learning. In the initial discussion in the class about the body, one boy said that the skin was to keep all the blood inside the body as otherwise it would leak out. After this lesson three children drew the body as a container filled with blood. It is however difficult to tell whether this comment had this specific effect or not or if these were merely the three children's own ideas.

When learning about the bones the children explored and counted each other's ribs by holding in their breath and making the ribs "visible"

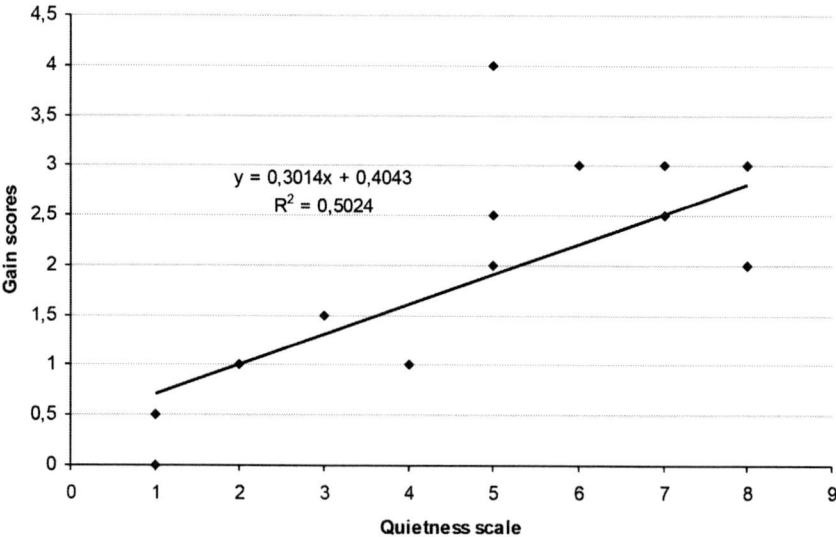

Figure 4.26 The scattergram shows the relationship between quietness and gains averaged for bones and organs (after the scales have been equated) and how knowledge increases with quietness.

and talking together about it. Here they were obviously learning from each other and together, and all drew the ribs on their drawing after the intervention, though this included much additional information. Another example was when they were working in pairs doing interactive tasks on the Internet, where they were required to put organs in the right place in the body and assemble a skeleton puzzle. This seems to have influenced their ideas, although it is again difficult to tell whether it was the activities themselves that had most effect or the interaction between the children, or both.

The effect Óli´s ideas had on the other children seem, however, most visible. He was enthusiastic and very keen on saying what he knew and the other children listened to him and took notice of him. There are several examples where Óli's ideas had a direct influence on the other children's ideas. To give one example, when talking about the brain, Óli said aloud in the class that the brain is divided into three parts (see Chapter 4.1). The teacher had decided not to talk about the different parts of the brain but this comment from Óli obliged her to talk about it and the many different "roles" the different places of the brain have. Óli had said that the three parts of the brain were the "main brain," the "small brain" that he also calls "control brain," and the third part that according to him is the "sleep brain." The teacher had a large photo of the brain from the *Let's Look at the*

Body (Óskarsdóttir, & Hermannsdóttir, 2001a), but the photo there does not show the different parts of the brain. So when explaining all the different areas of the brain she took another book, *That is How We Are* (*Svona erum við*) (Kaufman, 1976), that had a picture showing the different areas of the brain and told them about them. After this discussion and extra information from the teacher, the children were asked to draw a picture of the brain inside the head. The drawings of two children indicate strongly that they had listened more to Óli's ideas than to the explanations of the teacher because they drew and colored the brain in three different colors, though perhaps they may have just imitated each other's drawings as they sat near each other (Figure 4.27). Most of the children, however, drew the brain similarly to the brain in the picture in the big book, *Let's Look at the Body* (Óskarsdóttir, & Hermannsdóttir, 2001a), with the brain stem shown at one side of the brain.

This also shows the influence the children can have on the teaching and the course of the discussion. The teacher said, after the lesson on the brain, that if they had drawn the brain in three parts, this would probably have come from Óli. She also said that if Óli had not talked about the different parts of the brain she would not have talked about the brain in so much detail, commenting that this was an example of how the children influenced the route the teaching could take.

When I interviewed the children I asked them if someone in their class knew a lot about the body. Almost all of them mentioned Óli, and one child said that when the teacher had asked them, as an assignment, to talk about something in front of the class for five minutes the other day, Óli

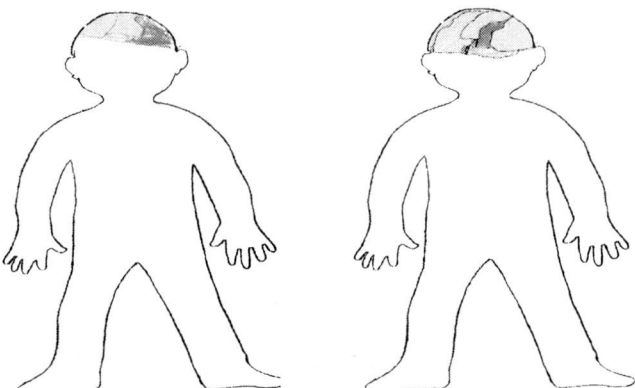

Figure 4.27 The drawings made by the two children who colored the brain in three colors.

had decided to talk about the body, and it was obvious that many of them were impressed by his knowledge. The teacher also said that Óli knew a lot about the body and had contributed a lot to the discussion, giving a lot of information to the class. She said she thought it important to give him the chance to say what he knew although she tried to involve the other children too. He was the one who usually came up with the right answers and the information and ideas that were important and relevant to the project. She said that sometimes she felt that the children listened more to him than to her. She was sure that the children learned a lot from each other: By sharing ideas as they were doing and working together they gave information to each other.

It is difficult to identify exactly which ideas expressed by the other children came from Óli and which from other sources; but there are clear signs to indicate that he had considerable influence on the other children's development of ideas.

After the lesson about the heart and the blood circulation the teacher pointed out to me that it was quite obvious that most of Óli's knowledge came from somewhere other than school. She said she had the feeling that Óli was not learning a lot because he knew so much already, but she hoped he was learning something and connecting together the ideas that he already had, and also stabilizing and "correcting" his ideas. "I think this has made him stronger as a pupil." His parents also said that the project about the body had enriched his self-esteem and they thought he had learned a lot from taking part in it even though he had known a lot about the body before.

5

Discussion and Conclusions

A number of different methods were used in order to answer the research questions that were put forward at the beginning of the study and will be revisited here about children's ideas and how these developed. Here the questions and the results will be reviewed in the context of other research, and the strengths and weakness of the approach assessed.

5.1 Children's Ideas About the Body—Structure, Location, Function, and Processes

The first research question is on the development and change in children's ideas:

> How do the ideas of 6–7-year-old children about the body change over the course of two school years during teaching about the human body in relation to location and structure (bones, muscles, heart, lungs, brain, digestive system), function (of the heart, skeleton, lungs, brain, stomach), and process (digestion and blood circulation)?

"The Brain Controls Everything", pages 155–188
Copyright © 2016 by Information Age Publishing
All rights of reproduction in any form reserved.

This question invited the sub-question:

> What kind of ideas do children in the 1st grade of primary school in Iceland (six years old) have about their body before teaching about the human body begins?

The study has shown that it is useful to distinguish between various aspects of knowledge and understanding, that is, the knowledge about structure and location of an organ or organ system and knowledge of processes and understanding of the function of an organ and how organs are interrelated (Cangelosi, 1992). This knowledge and understanding also varies according to the different organs, as children's ideas and knowledge about structure and location of the brain and heart are more developed than ideas about the structure and location of the stomach (see Figure 4.17, Panels a–d). This study provides significant information about the knowledge of structure, location, function, and processes, as some children seem to know about the structure and location of certain organs more than others at the beginning of the study, and more about the function of some organs but not so much about their structure and location.

The study also shows that there are two general issues that are noteworthy generally for the children at the beginning of the project; one concerns their misconceptions and the other the types of explanations they adopted initially but which gradually gave way to other types of explanations. The results show that the children in this study had at the beginning of the study similar ideas to those of children in other studies (Carey 1985; Osborne, Wadsworth, & Black, 1992). They are aware of their external body parts and they usually also know the organs like heart, brain, and stomach, as in other studies. There were also misconceptions similar to those found in Gellert's study (1962). In her study some children were not aware of the location of the lungs and thought they were located in the neck or in the head. In my study, a few of the children put the lungs in one of the shoulders on their drawings even after the teaching. In comparing children's ideas and knowledge about structure, location, function, and processes, they knew at the beginning of the study more about the function of the lungs than their structure and location, and nothing really about the processes involved. In Gellert's study some children thought the heart helped us breathe. This was also the case in this study when one child said the heart helps us breathe. It is perhaps not a straightforward misconception, as Gellert says, because the heart and the lungs work together and both heart rate and breathing rate increase when you run or exerclearned and decrease when you are relaxed. I assume that in these cases the children are trying to connect the bits and pieces of knowledge and understanding together even though they do not get the whole picture, something that I think is difficult and unrealistic to expect of children of this age. Later, though, they may be able to add to these bits and pieces and gradually get a better and more detailed perspective of the function of the different organs and the processes they are involved in. Sometimes it is difficult to tell just by listening to a discussion whether something is a misconception or not. In this case about the heart and the lungs I think it would have been helpful to follow this issue up and talk to the children in small groups or individually, and then take up the issue and explain to the child how the heart and lungs work together.

The children also used words and sentences that indicate they have "psychological ideas" (see Chapter 4.1) in contrast to "biological ideas" where they talk about their organs like independent "creatures" having wants and beliefs like those described by Carey (1985); these ideas seem to be based on "intentional causality," such as the heart has to beat and the stomach has to digest food, as if it were a matter of independent decision by these organs. However, I prefer the idea of a "vitalistic" explanation (Inagaki & Hatano, 1993; Slaughter & Lyons, 2003). These authors claim that children use a vitalistic explanatory framework when reasoning about the human body and when they refer to different body parts and organs as being independent and alive. "Vitalistic" implies the concept of being alive. There is a considerable overlap between the ways in which the terms "psychological ideas" and "vitalistic explanations" are used, but the strength of "vitalistic explanations" in this study is that the children seem to think and talk about the organs and different body parts as independent creatures. There are a number of examples in my study of what seem to be vitalistic explanations: "The brain controls everything," which indicates that the brain has a mind of its own and therefore this can be classified as a vitalistic explanation, albeit one about which philosophers might discourse endlessly. "The brain can crawl out if you get a hole on the head." Here, it is indicated that the brain is a creature that can crawl, so this comment can also be categorized as vitalistic. However, in my experience many grown-up Icelanders talk this way too when they are talking about the body, even though they know better, so perhaps we, adults and teachers, tend to retain these ideas into adulthood, as has also been found in other studies (Miller & Bartsch, 1997). But it may be the language rather than the ideas that is retained. It would be interesting to look into the influence the Icelandic language has on how children, and adults too, express their ideas about the body compared to other languages.

5.1.1 *Bones and Muscles*

Before the intervention, the children's ideas about their bones were similar to those of the children in the SPACE study (Osborne et al., 1992), and also similar to the results of Reiss and Tunnicliffe's study about children's knowledge of the human skeleton (1999a); as the children were more aware of the bones they could feel, such as those in the arms, legs, and ribs and also the skull. Their initial ideas were also similar to the results of a study undertaken by Guichard (1995) that shows 97% of children under eight years old do not see the skeleton as a functional structure. Just one child in this study (Child 5) scored at Level 6 on Reiss & Tunnicliffe (1999a) bones scale (see Figure 4.1), that is, 95% of the children in the study did not see the skeleton as a functional structure. The results of this study also show that the children's initial ideas about muscles were similar to the ideas of the children in the SPACE study (Osborne et al., 1992), where most children apparently only thought muscles were in the upper arms and possibly the legs. The children possibly have these ideas about muscles being in their upper arms from a very early age because one of the first things young children learn to do is to clap their hands and to show how strong they are by showing their muscles on their upper arms. The picture in the book *Let's Look at the Body* (Óskarsdóttir & Hermannsdóttir, 2001a) that goes with the text about muscles shows only the muscles in the upper arm (see p. 10), which also underpins this idea of muscles being in the upper arm. The question of imitation effects has to be taken into consideration as some of the post-teaching drawings suggested copying from the textbook, so that the drawings may be showing imitation rather than understanding.

After the teaching about the bones and muscles, and much discussion, exploration of models, pictures, exerclearneds with their own bodies, and different hands-on activities, the drawings showed that the children's ideas had changed substantially. After intervention, they were more aware of both the structure and location of the main bones (see Figure 4.1), with 35% of them now scoring at Level 6: Definite vertebrate skeletal organization (Reiss & Tunnicliffe, 1999a). Although it is not always clear whether something else might have extended their knowledge and understanding, this is unlikely here because of the short time between the pre- and post-drawings. It is difficult to see if ideas about the function of the bones changed, as this cannot be inferred from the drawings. From the other methods used to access the children's ideas, namely, classroom observation, interviews, and diagnostic tasks, it is also difficult to infer anything about children's ideas about the function of the bones, except they are aware the bones keep the body upright and the joints enable us to bend and stretch. According to their drawings made after teaching about bones and muscles, which included discussion about muscles being everywhere in the body and exercises that included bending and stretching, many children

(13 children) drew the muscles only on arms and legs (see Figure 4.4) and only 6 children drew muscles elsewhere in the body. This is in tune with results from the SPACE study (Osborne et al., 1992) although the intervention there seemed to improve more general awareness of muscles in other parts of the body than on arms and legs than in this study.

5.1.2 Organs

There was not so much change in the drawings of the (non-skeletal) organs between the pre- and post-drawings in Primary 1, and the teaching about the organs was not as detailed as the teaching of the skeleton, as the emphasis in Primary 1 in accordance with *National Curriculum Guide: Natural Science* (Menntamálaráðuneytið, 1999) was more on mentioning the main organs rather than giving detailed information about them, and the teaching only consisted of discussion and looking at pictures and doing activities on the Internet. The main reason for the small change between pre- and post-drawings of the organs can probably be explained by this. The drawings showed that many children did, however, know more organs after the lessons but generally did not know how these were related or connected. So, it can also be assumed here that the intervention in this phase in Primary 1 had an effect to a certain extent, but primarily in increased number of organs known by the children. These results are similar to the results of the SPACE study where the most significant change was the increase in the average number of organs drawn by each child after intervention (Osborne et al., 1992).

It was not until a few months later (in Primary 2) that emphasis was put on teaching the children about the functions of the different organs and how they were connected, following the aims and objectives in the *National Curriculum Guide: Natural Science* (Menntamálaráðuneytið, 1999). However, after teaching about the heart, lungs, and the blood circulation (Curricular Episode 7, see Table 3.1), many of the children were still quite confused about the processes that take place. Most of them knew that we use the lungs to breathe but could not say much more about them and their function. Some also seemed to confuse white blood cells and blood rich in carbon dioxide, which is not surprising as these are complex issues to much older children and even to adults. When drawing after the lesson they still drew a heart that was divided into two halves, one red and one blue, and started drawing red veins on one side of the body and blue on the other, though they changed this in the light of new information from the teacher who wanted them to make corrections. The drawings also showed that although most of the children put the lungs in almost the right position, some did not and did not show any signs of connection between heart, veins, and lungs, which suggests they did not see or understand the connection between the heart and the lungs, neither in positional nor in functional sense.

These results are similar to those of the SPACE research (Osborne et al., 1992) where attempt was made to gather information about the children's understanding of gaseous exchange. The ideas fell into three categories: in the first were ideas of everyday nature, such as the air we breathe goes into our tummy, keeps us alive, and well. In the second category, some greater knowledge was revealed about the organs' role in respiration showing an enhanced level of understanding. In the final category were ideas that were relatively rare but showed knowledge of gaseous exchange. The children that had those ideas were the only ones that drew the lungs on their drawings of what is inside your body (Osborne et al., 1992). Most of the children in my study knew, according to classroom observation and diagnostic tasks that we use the lungs to breathe although some did not say anything about this in the interview when asked about the function of the lungs and in that sense they fit into the first category of understanding of gaseous exchange (Osborne et al., 1992). There were only two children that mentioned oxygen and the blood getting oxygen from the lungs in the interview. These two children had however both drawn lungs (one only one lung) in their initial drawings of the organs, but as only these two showed any knowledge of gaseous exchange it is problematic to infer any correlation here between the content of the drawings and knowledge of the gaseous exchange process. Thus, the pattern obtained in this study fits as this issue is concerned with the pattern in the Osborn study (Osborne et al., 1992).

Even though the children seemed to be interested and paying attention to what the teacher was saying and demonstrating, many of them did not understand the process of blood circulation and the function of the organs involved. They had grasped some things but they seemed to be confused by other concepts. Both the blood circulation process and how the heart and the lungs work together seem to be too complicated for such young children. Their ideas are interesting nevertheless for those of us interested in teaching and learning. However, for those one or two who already knew a lot about the body and the function of the organs it made sense, and they were able to build up further knowledge and understanding or at least strengthen the ideas they already had, such as the blood received oxygen from the lungs. So, it is worthwhile for the teacher to talk about the blood circulation process and the organs involved, though it may not be of relevance for all the children.

In this case, about the connection between the heart and the lungs, I think it would have been helpful to follow this issue up and talk to the children in small groups or individually and then take up the issue and explain to the child how the heart and lungs work together.

This information should be of special value to the teacher in deciding what to teach and which teaching methods to use, and thus should be a valuable contribution to educational practice.

It is also interesting that in the SPACE study, Osborne et al. (1992) found that some individuals in the middle age group (i.e., lower juniors [8–9 years old]), showed a level of understanding that indicated some knowledge of circulatory process which some individuals in the upper juniors did not show, although many lower and upper juniors mentioned veins. Perhaps children are reluctant to express their ideas as they get older because they are afraid of saying something wrong; they may also have many ideas but cannot put the concepts together because of something missing and consequently cannot get the whole picture and therefore decide not to express or share their ideas. Thus, it may be that the lower juniors in the SPACE study were more open to expressing their ideas than the older ones but not necessarily that they knew more.

As the results show, the children's ideas about the organs did not change greatly between pre- and post-drawings, although they knew the names and structures and positions of more organs, they generally were aware of their major functions but in a very simple way, like "the heart pumps blood," "we breathe with the lungs," and "think with the brain." However, the following year, after the curriculum episode about the heart, lungs, and blood circulation, the children showed in their drawings that they knew the blood from the heart went into veins and traveled through the veins all around the body. Knowledge of the function of the lungs in the process of the blood circulation, as shown in class discussion and interviews was vague, although many realized there is a connection between the heart and the lungs—for example, that you breathe faster when your heart beats faster. Most of the children who drew a heart divided into two halves, one red and one blue, were obviously imitating the picture on page 14 in *Let's Look at the Body* (Óskarsdóttir & Hermannsdóttir, 2001a). Representing the heart like this in a drawing is more likely to be a symbol for the heart than a realistic picture of it. According to class discussion, interviews, and diagnostic tasks, the children in the study generally knew that the heart was essential for living and they used comments like "the heart beats" and "it pumps blood into our body" in terms that explain its function or rather its behavior.

Only four children drew lungs in their initial drawings and none drew the liver or kidneys. These words are not a part of our daily language, as opposed to the heart, stomach, and the brain. We talk about the heart that beats fast and also describe emotions like "I love you with all my heart." We talk about the tummy, full tummy, sore tummy, and empty tummy when we are hungry. We talk about using our brain to think and find out things, and when telling someone off or helping someone to find something out we sometimes say: "Use your brain"!

Just a few years ago lamb livers and kidneys were seen on Icelandic dining tables much more often than nowadays. To a certain extent this is still so in the countryside and it would be interesting to see if children from the countryside are more familiar with the structure and the function of the liver and the kidneys than children in Reykjavík.

5.1.3 Digestion

The teacher in this study thought is was important to just "talk" about digestion in Primary 1 (Curricular Episode 3) and get to know the children's ideas through discussion and drawings, but not to teach them about the process and all the different organs involved until they were in Primary 2 (Curricular Episode 8, see Table 3.1.), as recommended in the *National Curriculum Guide: Natural Science* (Menntamálaráðuneytið, 1999). In their initial drawings, only four children drew food in the stomach mixed or digested; all the others drew the food in whole pieces in the stomach. When asked if they really thought they could swallow a whole apple or a carrot they said they did not, but just drew it this way because they did not know how to draw it otherwlearned. This is important as their competence to draw determines how fully they can express their knowledge. The children knew that you have to chew food to be able to swallow it, and so it seems that it is their natural way to use a symbol, but it has to be borne in mind that this does not really show their knowledge. If their drawings had been followed up individually by asking each child to explain their drawing, it is likely that a different and a fuller view of their ideas would have been attained. So, here the drawings do not tell the whole truth about their knowledge and understanding. This is similar to the results presented in the SPACE study where some children drew food in all parts of the body and explained it by saying that the food went into the blood and the blood went around the body (Osborne et al., 1992). In that study, the drawings drew a very symbolic picture and the food drawn all around the body was a symbol for the nutrition from the food, and shows a certain degree of understanding.

The children in both the present study and in other studies (Driver, Squires, Rushworth, & Wood-Robinson, 1994; Osborne et al., 1992) think that the stomach is much bigger than it really is, but they know roughly where it is located although generally thinking it is lower than it actually is. Here in Iceland we talk about "the tummy" as the whole middle area from hips to chest, and so it is no wonder that the children find the stomach very small in reality and do not recognize its structure when they see it in a torso or on a picture, but instead get confused and mix it up with something else. This was also the case in the SPACE study where many of the youngest children simply drew the stomach as a big bag that contained untransformed food, and also in Carey's study (1985). By the age of seven the children start to realize that the stomach helps to break down or digest food, but it is not until later that children understand that food is transferred elsewhere after being in the stomach (Carey, 1985; Driver et al., 1994). According to Carey, very few 9–11-year-old children knew that food changed in the stomach and broke down into altered substances that went to tissues throughout the body. All the children in my study knew before the project started, like the children in Carey's study (1985), that you need food to grow and be healthy and to be strong and that you die if you do not get anything to eat. Their initial general ideas about the function of the stomach all related to digestion like: "the food digests in the stomach" and "the food gets mixed in the stomach."

The 4- and 5-year-old children in Toyama's study (2000) said that food went to various parts of the body, indicating they seemed to have sufficient insight to accept some material transformation of food. As my results show, all the children changed their drawings after the demonstration with the Weetabix and milk and the explanations and discussions that followed. But perhaps the demonstration helped them clarify their ideas and present and show their ideas better in a drawing. There were, however, only two children who mentioned in the interview something about nutrition going from the mixture in the stomach to the rest of the body. Their ideas were, however, very unclear and their knowledge and understanding very vague, but they still knew something about it, though not enough to be able to explain the process in much detail in the interview. This was also the case in the study of Carvalho, Silva, Lima, & Coquet (2004), but according to their study the drawings the children made showed a great deal of confusion about what happened beyond the stomach. This was despite the fact that the children could describe orally the correct sequence, which suggests they had learned it by heart although their understanding may have been vague.

Because of the difference between the post- and pre-drawings of the food in the stomach, one can conclude that the teaching that took place between the drawings had a considerable effect on children's ideas, and certainly helped them to clarify their ideas. The drawings, however, can still be symbols for their ideas or an imitation, and their ideas may not have changed so much after all, because they said after they had drawn the food in whole pieces that they knew they could not swallow it like that, even though they had drawn it that way. It was also difficult to get them to explain the digestive process in the interview but most of them did point out the organs involved in the digestive process and the way the food went from mouth to anus. So at least the demonstration changed the way they represented food in the stomach and the teaching about digestion made them able to get a picture of the way the food goes and the organs involved so at the end of the project they knew more or less the location of the organs that make up the digestive system. At least they could point to all the organs involved in the interview and show the way the food goes from mouth to anus, even though they could not mention the names of all the organs, and many of them were aware of the function of the stomach and also the digestive process, even though they did not describe it in detail.

5.1.4 Brain

It is interesting how few studies discuss children's ideas about the brain because in this study the children seemed to think it was a very interesting organ. Furthermore, they all knew something about the brain although only three boys expressed their ideas in class discussion—rather, it was as if they were telling the others or teaching them all they knew about the brain. Thus, most of the information about the brain came from these three boys, especially from one of them, Óli (5). The teacher added a few things in the light of their information and all the children did were a few exerclearneds involving the senses.

Although in the interview the majority of the children seemed to know about the structure, location, and some of the functions of the brain, two of them did not show any knowledge of the function of the brain in the diagnostic tasks. It is quite possible that these two could not read or understand the questions and statements about the brain and did not ask for help or get the teacher to clarify. Only three children, according to the diagnostic tasks, knew that nerve cells bring messages to the brain and back. In the light of the exerclearneds the children did about their senses, and the drama the teacher did with the girls about the nerve cells sending messages to the brain and all the discussions involved, it is somewhat surprising that only three children could answer the question and the statement about the function of nerve cells. Perhaps the diagnostic tasks were too difficult for some of the children; they had to concentrate and complete the task and may have felt insecure by doing this on their own. The teacher and I helped the ones we thought needed help reading the text. The children were not used to taking tests and having to complete a task like this, and so this might also be the reason for some not completing the task rather than that they didn't know the answers. This ralearneds a recurring issue. Using just one method to get information about the children's ideas does not give the right, or at least not a consistent, picture of their ideas, or at least not the whole picture. On the basis of the present results, it may be inferred that these boys, especially Óli, had an effect on the ideas that the other children developed about the brain. The textbook had also a considerable effect and it is obvious that the children had the drawings of the brain in the book in mind when they did their own drawings of the brain. However, from the results, it can be concluded that teaching about the brain and its importance and functions can be very suitable for young children as the brain was one of the few organs they knew and drew in their initial drawing of the organs of the body.

To sum up, the children's ideas about the bones and the skeleton changed more than their ideas about other organs according to the drawings, and the results show that the variety of activities and teaching methods used had significant effect. The drawings of the food in the stomach changed also in their drawings, but classroom observation of class discussion indicates that the ideas perhaps did not change so much after all but rather it was their way of showing it that changed. Their drawings of the heart, lungs, and blood circulation did not show much difference, as many still drew Valentine-shaped blue and red hearts even though they knew the heart did not look like this. So there are different kinds of information represented in the drawings and it is hard to see just from the drawings the ideas the children have. The children's fine motor skills also vary greatly. The drawings are, however, of great value, and asking children to draw their ideas is an important method to gain access to children's ideas, but it has to be used along with other methods and the children have to be encouraged to talk about and explain their drawings.

In Reiss and Tunnicliffe's study (2001) the digestive system was one of two organs systems best represented. The children in this study were also more aware of the digestive system than other organ systems (organ system where organs are interrelated and make a "system"), and more aware of the digestive process than the other processes, possibly because digestion is a process that is so "visible," or at least the beginning and the end of it! In this respect the results of this study are similar to those of other studies, except that no other studies of children's ideas about the brain were found except once where the brain was mentioned when talking about children's ideas generally. This study shows that the children were aware of the location and main function of the brain although at a very simple level. What is also important here is the change in children's ideas represented in their drawings, which leads us to the influencing factors (discussed in the Section 5.2). It is also interesting to see what does not change in their drawings, such as the shape and the color of the heart and the size and the structure of the stomach, even though their ideas have changed and developed.

At the end of the project, the children were generally more aware of the structure, location, and the function of the different organs than they were of processes and how the organs were interrelated. In light of this, it can be concluded that it is important to distinguish between the structure of an organ, location, and function and then teach about processes of specific organs and organ systems as recommended by Reiss and Tunnicliffe (2001).

5.2 The Main Factors Influencing Changes in Pupils' Ideas

The second research question focuses on the main factors influencing the changes in pupils' ideas:

> What are the main factors influencing the changes in pupils' ideas: the curriculum, teaching methods, teaching materials, teacher–pupil and peer interactions, or something else in addition to these? How are changes in pupils' ideas affected by: the curriculum, teaching methods, teaching materials, teacher–pupil and peer interactions, or other factors?

This question provided much information as described in Chapter 4 (Results), mainly through classroom observation and video recordings, but also through looking through children's drawings to see if it was possible to connect changes in ideas as put forward in a drawing, with specific teaching methods and/or the teaching material, or with interaction in the classroom.

5.3 Teaching Methods: Which is Most Effective?

The preceding discussion suggests that it seems easier to teach about some organs or organ systems than others; perhaps children are more interested in some organs but not so interested in others; or some teaching methods are more effective than others for some children but not for all. The teacher in the research was deliberately trying to build on the children's existing ideas and trying to extend their knowledge and understanding by using different teaching methods. In that sense it can be suggested that she was putting her effort into the "Zone of Proximal Development" (ZPD), and using different types of scaffolding to support their learning (Ormrod, 1995; Vygotsky, 1978; Wood, Bruner, & Ross, 1976).

The teacher's way of getting the children's attention was interesting and something that was not pre-planned. If the children were not listening or not taking notice of what she was saying, she looked at them and waited for their attention. The teacher also used eye contact to try to encourage the quiet children to make oral contributions. Thus, eye contact and gestures were also important (as discussed by Hayes, 2004). The quiet children obviously did not, however, get much encouragement or positive comments about their contributions to the discussion as they did not share their ideas and thoughts.

I classify the teaching methods used into two categories, depending on the aim of the teacher. These two categories of teaching methods overlap and support each other. The first, aims at determining the present status of the child's ideas, that is, the discussion method and questioning strategies (question-and-answer method) where the teacher asks some key questions of the children. The second category is methods that aim at extending the children's ideas (e.g., an introduction, mini-lectures, practical work— hands-on activities or investigations, interactive activities on the Internet, and drama and demonstration). Discussion can also extend and develop ideas, and both the children that share their ideas and the others who only listen can benefit from the discussion. Listening to children working on practical activities or doing interactive activities on the Internet that aim at extending their ideas can also give the teacher important information about children's ideas, and all of this can and should be looked at as an important part of formative assessment to be used for building on children's experiences and knowledge, and extending their ideas and concept development.

It is difficult to tell whether one teaching method is more successful than others and in what sense it is successful. There are so many things that have to be taken into consideration and sometimes more than one method is used at the same time. One child can learn a lot by listening to the teacher or a classmate; another child can clarify and extend her or his own ideas by talking about them. Feedback from the teacher is important for the child and the feedback can sometimes be just in the form of a mime and gestures of the teacher and still have effect, both encouraging and discouraging (as discussed by Hayes, 2004). Thus, a variety of methods should be used in order to ensure rich understanding. It is also clear that a variety of other methods, such as oral introduction, pictorial presentation, and demonstration, may have considerable impact, but they must be used in conjunction with other methods, as discussed by Kozulin (2003) who says that by a symbolic tool like a map of Italy including Rome, when teaching about Rome, can help the students apply, memorize, and understand better the content knowledge. This would also apply to using a picture or a model of the heart or the skeleton when teaching about the human body.

Demonstration and discussion at the same time seems to be very effective with this age group (6 and 7 years old), and seems to be even better than hands-on activities that the children are doing on their own where it can be difficult for the teacher to follow what is going on or the discussion that takes place. However, practical activities where the teacher is watching and listening to the children or leading the activity like the activities about the bones, muscles, and joints and the activities using the stethoscope to listen to the heartbeat seem to have a significant effect, though the discussion taking place at the same time could also have had an effect. The practical tasks that involved all the different activities concerning the bones, muscles, and joints seemed to have a considerable effect, as results in Chapter 4.1 and 4.2 show.

The findings of this study support Osborne (1996) when he says that the advocates of constructivist methods of teaching have failed to recognize that there is a role for telling, showing, and demonstrating. Instead, they have chosen to emphasize cooperation and discussion-based activities in order to promote knowledge, ignoring the fact that there could be a place for telling, showing, and demonstrating and the possibility that what constructivist pedagogy offers to learning is preferred by some students but not all, and consequently is effective for some students but not all. Therefore, there is no single method for teaching and learning that fits all; so science education should consist of a wide variety of teaching methods, including telling and showing (demonstrating), and take into account the truth that each individual learner is unique.

The teacher in my study used methods adopted from Piaget when she usually began each lesson or a new topic by encouraging the children to express their ideas and then she followed up with prompts and challenges in relation to their ideas (Noddings, 1995). She also used methods adapted by Vygotsky, when she guided and led the children by interacting and talking to them (Hodson & Hodson, 1998). The social constructivist view suggests that the most effective form of learning is inquiry oriented, personalized, and collaborative which means the child can both work alone and in collaboration with others (Hodson & Hodson, 1998). The teacher in the study obviously had this in mind, even though she had not thought especially about the theoretical background for the methods she used, as they were just natural to her, built on her beliefs and experience. But by using the discussion method, she said she felt that she could hold children's attention and they seemed to listen and be interested in what she said, and this supported her in using it. She said she felt that when she was using the discussion method she was in control and it made her feel good and confident and the teaching easy. The results suggest that this was a good way to get the children interested and hold their attention. However, it is not so easy to see whether or not the children have grasped the issues and concepts being discussed, because so few children took part in the discussion, and therefore, the teacher got the ideas from only a small sample of the class. Perhaps this sample reflects the ideas the other children have but it is impossible to tell only from using class discussion.

The results of the study show that lecturing and telling can work well for some pupils in some cases, especially if the teacher uses something visual, like models and pictures to clarify things, or even employs just a change of voice and puts on a "little act" as the teacher in the study did when teaching about the heart and the blood circulation in order to build on further learning opportunities and facilitate the pupils' learning. It is, however, difficult to tell exactly what the children learned exclusively from the drama because drama was used together with other methods, but from the interviews and diagnostic tasks, it did not seem to have had an effect though it could have influenced and clarified their understanding.

The results show that different teaching methods have different effects on different children. Thus, a variety of methods should be used in order to ensure a rich understanding and to take ideas further. What might be called "passive" methods, such as oral introduction, telling, pictorial presentation, and demonstration, may have considerable and perhaps unique impact, but they must be used in conjunction with other methods. Similarly, teachers should use a variety of methods to obtain access to children's ideas and not make assumptions on the basis of only one method. If we are encouraging teachers to adopt the constructivist view of learning and teaching, class discussion is not enough to find out children's ideas to build further work on; other methods have to be used too. Discussion (both class and individual discussion), drawings, and practical activities and tasks should give the teacher access to children's ideas and be a part of formative assessment about certain issues and topics to help her plan a further curriculum that builds on children's existing ideas, and aims at extending them further. Thus, the different teaching methods the teacher used played an important part in the scaffolding process.

The results suggest, however, that the teaching had more influence on the quiet and semi-active children than the active children, as the quiet children and the semi-active gained more than the active children. So, the teaching methods and the teaching material used perhaps did not suit the active children as well as the others. When related to Vygotsky's ideas about the ZPD it can be suggested that the methods used extended the ZPD for the quiet and semi-active children, especially in relation to the bones, but not as much for the active ones.

The teaching phase and the methods used seems to have great influence on the change in children's ideas as the majority of the children in the class added to their knowledge. All the children got the same information and the teaching methods used seem to have worked for the children as a group as they all improved up to a similar level. So perhaps the teaching itself played a greater role than their activity level. This begs the question whether the ones that did not add to their knowledge may have needed something additional that aimed at extending their ideas.

5.4 The Influence of the Teaching Material

The teaching material used, *Let's Look at the Body* (Óskarsdóttir & Hermannsdóttir, 2001a), clearly had an effect on the children's drawings because they tended to imitate the pictures in the book when they were asked to draw their ideas. There is a big picture in the book that is supposed to illustrate the blood circulation system. The picture shows the heart shaped like a Valentine's heart and has red and blue veins around the body. This picture seems to have had a great effect on many of the children because they drew the heart quite like the heart in that picture in the book, half blue and half red, even though the teacher had shown them a model of a real heart and talked about the heart and how it does not look like a Valentine's heart in reality.

A picture that goes with the text about the digestion system shows the mouth full of food in whole pieces: an apple, a carrot, a bread slice with cheese, etc. In their first drawing of food in the stomach the children drew the food like that shown in the picture in the book, like whole apples and carrots even though they knew that you would choke if you swallowed the food in whole pieces. There are two similar pictures of the brain in the book. Both of them show a brain with the brain stem coming down from one side of the brain. When the children drew the brain, many of them drew it like the brain in the book with the brain stem coming down from one side.

The effect of the drawings in the textbook on the children's thinking should not be undervalued as they seem to have more effect than might be expected, as the children imitated the drawings in the book even though the drawings did not represent their ideas. They drew the food in whole pieces in the stomach even though they knew better; they drew the heart in two halves, red and blue, even though they knew the heart was not exactly like that; they drew the muscles in the upper arm as the picture in the book shows; and many of them drew the brain exactly like the drawing of the brain in the textbook. This was also the case in the study of Carvalho et al. (2004) where the great majority of year 3 and 4 children reproduced drawings from the school textbook about the digestive tract.

Teachers' use of metaphors and analogies while explaining scientific things has been discussed in the literature (Holgersson, 2003; Ogborn, Kress, Martins, & McGillicuddy, 1996) and it has been argued that metaphors and analogies play an important role in explaining scientific issues. In the text book *Let's Look at the Body* (Óskarsdóttir & Hermannsdóttir, 2001a), certain words that children are familiar with are used to illustrate functions and processes such as "guard dogs" for white blood cells and "a mixer" for the stomach; the blood travels through the veins like "trains" and the liver is the "cleaning factory." The teacher used these words or metaphors when explaining the organs and the functions and processes that they were involved in. These words or metaphors also seemed to have an effect as borne out in the interviews where some children remembered them and used them to illustrate their ideas. However, most of them did not seem to understand what these words really stood for even though they remembered them and tried to use them. But, I still think they helped some of them make connections. They also used metaphors like "the heart being a pump." Adults use metaphors in daily life, like "the heart wants to be noticed" when the heart is beating fast, "my head is about to burst," "my tummy is bursting," and "my brain is full of ideas." These metaphors adults use are likely to influence children in using them as well but it does not mean that these will lead to and maintain misconceptions.

5.5   The Potential Effect of Interaction in the Classroom, Teacher–Pupil, Peer Interaction

The constructivist view of learning and teaching has guided me through this research where the emphasis is on interaction in all possible forms as a precondition for learning. I realize, however, that the constructivist view has its limitations. I agree with Cornu, Peters, & Collins, (2003) when they say that constructivism is in action when and where learners are active in the process of taking in information and building knowledge and understanding, that is, constructing their own learning. According to the study, the 19/20 children in the class are quite different. They are different individuals and their way of learning is different, both regarding the pace of learning and style. *The National Curriculum Guide: Natural Science* (Menntamálaráðuneytið, 1999) implores teachers to take children's conceptions and ideas into account when planning the curriculum, and the aims and objectives, skills, and attitudes at each stage build on, and add to, the aims and objectives in the earlier stages.

Taking children's conceptions and ideas into account when planning a curriculum can, however, be complicated because of how different the children are in their ideas and their individual learning styles. There is no one right way or one prescription to follow, although the constructivist view emphasizes the active engagement of pupils in their own learning, and pupils' engagement in action, discussion, and reflection gives the teacher opportunities to know their ideas and their thinking in order to facilitate their learning (Noddings, 1995).

The results indicate that working with peers, and information from peers, can have a considerable effect on the development of children's ideas; indeed, in some cases peers have no less influence than does information from the teacher or from the text in the teaching material. In the discussion about how babies are made, the children sort of guided each other through the discussion with the help of the teacher who asked questions. In this case, the discussion went from ideas about God creating us and the theory of evolution, to ones about a seed in the man that went into the woman and this would make a baby. Here, the teacher was clear about what she was looking for, but she did not say that any of the other ideas were wrong or comment on them in any way. She just asked new questions that led to the fact that the baby is made from a cell from the father and the mother. So, in this case, the teacher was a guide or a leader but the peers and classmates made the contributions and extended the issue further. Therefore, it has to be stressed once more that working with peers and sharing ideas is essential and whole class discussion is important, even though not all the children take part or share ideas. The children who share and talk, and the teacher's comments, questions, and explanations can have a great influence in helping all the children construct their ideas, knowledge, and understanding.

5.6 The Difference Between the Quiet Children and the More Open Children

According to the constructivist view, social interaction more or less builds on the view that conversations between children and adults are crucial for cognitive development and this view is emphasized in the literature

(Fosnot, 1996; Vygotsky, 1986). The constructivist view emphasizes also the importance of social interactions with peers and classmates where children are presented with ideas and points of view that either fit or differ from their own, and to which their existing ideas are forced to re-equilibrate (Ogden, 2000).

As it became obvious early in the study that there was a group of children who were reluctant to contribute to class discussion, the last research question has a rather special focus in the research:

> What are the differences between quiet children and more open children in respect to these issues (the influencing factors)?

This last question opened up a whole new area as it gave information about how children of the same age, in one class, can be very different from one another—how their ideas, knowledge, and understanding can be different, and how they learn at a different pace and use different methods to learn. In this study, the attention was directed to one aspect of these differences, that is, the visible activity (or passivity) of the children. This is particularly important as it relates to the theoretical background of the research to the extent that it is based on constructivist ideas of learning.

As the constructivist view emphasizes the importance of social interaction on cognitive development, it is important to involve quiet children in discussions as they benefit from listening to the ideas of the other children and the questions the teacher asks, even though they do not share their ideas. The quiet children in this study did not learn less than the rest of the children and they have nothing in common other than being quiet. In my view, the social constructivist view of teaching and learning does not put enough emphasis on this group of children, but as this study shows, quiet children can be one-third of the children in a class, which fits well to my experience as a former primary school teacher. Quiet children may be active by listening, looking, trying out things, making drawings, and completing tasks, and they can be interacting in their way without taking part in the discussion, but we do not know this if we do not put an effort into finding out their knowledge and understanding by using different methods. However, it is still the discussion that plays such a big role in the constructivist view of learning and to be able to build teaching on children's ideas we have to get access to them and try to involve all the children, including the quiet ones.

Early in the research I clarified the phrase "being active" as being active in discussion, that is, taking part in the discussion and sharing ideas. Both the teacher and I used the word "active" in a very narrow sense at the beginning of the study, namely describing children who were visibly active although both of us knew from a long teaching experience that being active in a classroom situation could mean a lot more than just taking part in the discussion and sharing ideas.

According to my research, a child can be an active listener and learn even though the child is not active in sharing ideas. Children can also learn by watching a demonstration, pictures, models, or a video, and they can be active in different ways, depending on the activity, for example whether taking part in a discussion, listening to the views of a peer, or watching a demonstration. But taking part in the discussion seems to be commonly used as a criterion when being active in the classroom is discussed (Collins, 1996; Myhill & Brackley, 2004). There also seems to be strong tendency in the discussion in the literature about activity in the classroom to assume that quiet children are not learning as much as the children that take part in the discussion. Results from this study, however, show this does not have to be the case as the children in the visibly passive group were definitely not learning less than the others, perhaps learning more. According to their pre- and post-drawings, they were learning (see Figure 4.22 and 4.26). Therefore, it has to be very clear what "being active" means in any discussion. In this research, "being active" in the classroom means they are learning and I normally assume that learning implies cognitive activity. It can mean an active listener or an active watcher. It can mean active interaction within oneself, with the environment, or with the things being explored. But, whether or not a child is "active," does not necessarily have to be visible. So, it can be assumed, as pointed out by Cornu et al. (2003), when and where learners are active in the process of taking in information and building knowledge and understanding—constructing their own learning—that is where we find constructivism in action.

Using the discussion method that is emphasized by the constructivist view, is not enough in a whole class setting because then the teacher just gets the view and ideas from just a small sample of children, the verbal children, while the quiet children do not normally share their views and ideas. Discussion is, however, very important; the open children get even more confident and the quiet children get to know the ideas of the others and learn from these too. An example of this is the class discussion about the brain, where Óli and two other boys shared their knowledge with the rest very confidently. It is important to encourage them to take part in the discussion, especially in small groups where they feel confident; like in the lesson when the children were learning about the senses and the teacher had divided them into two groups by gender and then into smaller groups (of 3 and 4). In this lesson, the children generally participated more and expressed themselves more than usual.

If there is a small group of children that share their ideas, as was the case in this study, it can be difficult to see if the children have grasped or not grasped the issues and concepts being discussed, and the ideas presented are just ideas from a small sample of the class. Perhaps this sample reflects the ideas the other children have but it is impossible to tell just by using the discussion method.

This study demonstrates that some children are willing to participate, and some are not. We just hear the ideas the *visibly active* children hold. The question about the quiet passive children in the classroom and the dilemma in constructivist teaching between serving the needs of the individual child and the entire class, has been ralearnedd by other studies (Collins, 1996; Keogh & Naylor, 1996; McCroskey, 1980; Osborne, 1997). According to this study, the *visibly active* children did not learn more than the other children, which is a key finding. But on the other hand, their ideas do not progress as much as the ideas of the others perhaps because they score higher on their pre-drawings of the bones and the organs (see Figures 4.24 and 4.26). This indicates that their preconceptions were more developed when the project started and the intervention that took place between pre- and post-drawings did not, according to the drawings, extend their ideas very much. They knew a considerable amount initially and the teaching did not really add to that very much. There are just 4 (5) children in the visibly active group so it is difficult to draw firm conclusions in the light of the results. It is however interesting that the children that knew quite a lot at the beginning of the project did not develop and extend their ideas and knowledge as much as the others (a ceiling effect explanation was discounted). The visibly passive children, the quiet group, and the semi active group were very similar in terms of their final state of knowledge and understanding. The quiet children had at least as advanced knowledge and understanding as the semi active children. This is also a key finding.

When the three groups of children are compared, that is, the visibly active, the semi active, and the visibly passive (quiet) children, similar progression is found in the semi active and the visibly passive group. But the visibly active group changes differently from the rest. Their gain is not as great as shown by the others, but their initial scoring is higher, which indicates that they knew more than the others at the beginning of the project. They however do not seem to learn a lot from the intervention, not as much as the other children, which means they perhaps knew most of it before the intervention occurred. This is a major issue that needs to be explored further. In the class discussions only a small group of children are usually willing to share ideas—the *v*isibly active children. These are the ideas that the teacher uses to build the teaching on and organizes the curriculum in the light of these ideas, but according to the results, the children that come up with these ideas learn less or rather add less to their ideas than the others. They are perhaps so enthusiastic to express their ideas and so eager to talk they do not take in what other people say, that is, their activity actually hinders their learning. Perhaps they get such a positive response from the teacher and the rest of the children that they do not feel they have to learn more so they are perhaps not motivated enough. Or perhaps they do not get an opportunity to learn more as the teaching is more or less aiming all at whole class teaching, or at least the same outcome for all the children. It has however, to be borne in mind that the children in the visibly active group are so few, so that even though the interaction pattern is statistically significant, one should refrain from drawing strong conclusions.

Collins (1996) emphasizes that "invisible" children have to be encouraged to be more assertive and find their voice in the classroom while the more vocal pupils have to be persuaded to talk less and give the quiet children a chance. Collins (1998) goes as far as to describe the pupils that are exhibiting quiet and non-participatory behavior in class as "playing truant in mind." I find it difficult to share that view on the basis of the data as a general assertion, even though this may be true in some cases. But still, these pupils are not doing what the teacher expects them to do and not taking part when and where they are supposed to so, in that sense, they are not following curriculum instructions and aims.

According to Collins (1996), "being invisible" does not, however, have to have a significant effect on children's learning and the results of this study show that the quiet children are not learning less than the others, they are even in some cases perhaps learning more (see Figure 4.22). The results also indicate that the quiet children do not seem to have more in common other than being quiet; beyond that, they are as different from one another as any other human being (McCroskey 1980). But what does this suggest? That social interaction in the classroom and class discussion is not so good after all, and does not seem to play the anticipated role in learning? What about the ideas of social constructivism that social interaction and sharing of ideas is important if not essential for learning? Is it then maybe better just to sit and be quiet? The results of this study do not suggest that, even though the results show that the quiet children are not learning less than the other and that quietness is correlated with gain as supported by the results (see Figure 4.26). In light of this, it can rather be suggested that their ideas developed because they listened to the discussion of information from the teacher and pupils, that is, sharing of ideas, and therefore got opportunities to relate new information to prior ideas and knowledge, possibly by reflective self interaction.

Sometimes children are reluctant in expressing their ideas because they think they are wrong and are afraid of the attention their comment might get. Therefore, the atmosphere in the classroom must be such that the children feel free to express their ideas even though they can be wrong. The teacher in my study tried to create that kind of atmosphere in the classroom. She asked open questions and encouraged the children to take part in the class discussion. She tried to involve all the children, including the quiet ones, and was aware of their reluctance to contribute. She did not say if an answer or a comment was wrong, at least not until she had got a good sample of ideas. Then, she usually summed up the ideas that had been put forward and tried to clarify unclear issues by telling and explaining. All ideas put forward were respected, but if they were wrong, the teacher tried to correct them again by telling and explaining and using pictures and models if possible.

Osborne (1997) and McCroskey (1980) both suggest that participation by thinking out loud should not be forced on the child and that teachers should not try to change the child's personality but instead take positive steps towards trying to involve the quiet child in the discussion. In a whole class discussion, a few children usually come up with a sample of ideas and the teacher uses these to decide what ideas to build on and organizes the curriculum in light of these. This was the case in this present study, but drawings were also used by the teacher to investigate children's ideas. To the extent it is sensible to encourage teachers to use the constructivist approach to learning and teaching, class discussion does not suffice to find out children's ideas. Other methods include discussion in small groups, drawings, concept maps, concept cartoons, and ideally, peer or individual interviews in some cases if possible.

Osborne (1997) also talks about children that dominate the class discussion, that know a lot and are eager to share their view, but sometimes at the expense of the other children who then do not get the opportunity to share their views. In the present study, there were two boys in the visibly active group, who were very eager to share their ideas and needed a lot of attention, but did not really have a lot to contribute to the discussion. Another boy, also in the active group, was however, quite different because all his contributions were relevant to the discussion and there are a number of instances where his ideas can be seen to have direct influence on the other children's ideas. In order to involve the other children, the teacher sometimes let him wait with his hand up for some time while she asked the other children to make contributions before she called on him. The children obviously respected his ideas and knowledge and the teacher was perhaps afraid that the children would be more reluctant to share their ideas if he had already shared his. This is also borne out when the teacher says that she sometimes thinks the children are listening more to him than to her. She was also worried because she thought he was not learning a lot about the body as he had known it all before but she hoped that he was at least connecting and consolidating his existing ideas. The boy also realized that he was not learning a lot but still consolidating his ideas and as he himself said he was "learning more about the things he already knew, and knew now better the names of the organs." According to his parents and also the teacher, his self-esteem had been enriched because of the project since he found that his contributions were respected and valued both by the other pupils and by the teacher. He had ever since he was a little boy (a toddler) been very interested in everything that has to do with science, nature, and technology. He was especially interested in the complexity of the human brain and knew even more than the teacher and the researcher about the brain, so he got a lot of positive and encouraging mime and eye contact from both of us. Some would argue that he should have been given something else to work on since he knew so much about the body—something that extended his knowledge even further. In this particular case, I think it was not necessary, even though in other cases it might have been appropriate—it all depends on the children concerned, but also the circumstances. This topic was exactly what he was interested in and he blossomed as a pupil in these lessons by the attention his contribution got, but according to the teacher this was not necessarily the case in other lessons where the topics were different.

Another, but a quite different, example is a boy in the semi-active group that also sometimes had difficulties with concentration and had fine-motor skills that were not as developed as many of the other children's. Because of this, it was difficult to analyze his drawings. He had difficulties paying attention in the class discussion and difficulties in concentrating and doing the diagnostic tasks. However, he blossomed in the interview and presented a totally different image of himself and of his knowledge and understanding than he had done in the classroom. Perhaps this child would be able to concentrate and work better in a small group of children, or even on his own—doing individual tasks.

There were also gender differences. The difference between the engagement of boys and girls in the class discussion in this study was remarkable. There were three to four boys who took an active part in the discussion (Level 1 and 2 on the quietness scale, Table 4.1), but only one girl (Level 3 on the quietness scale). Furthermore, she did not share her ideas as often as the boys that took an active part. Three other girls occasionally took part in whole class discussions, adding a word or two, but they usually did not take as active a part as the four boys did. This was also the case with some of the boys that just occasionally added in a word or a comment but did not really share ideas. The girls in general were also generally quieter and more passive according to classroom observation. I learned the names of the boys much faster than those of the girls because they were much more "visible" and some of the boys had difficulties in following instructions from the teacher. When the class was divided into groups by gender, all of them (both boys and girls) took a more active part in everything, although the girls were still more passive than the boys who took more initiative in the discussion and the activities, whereas the girls always waited for the teacher to give instructions and lead the work.

At the beginning of the study, I found these gender differences surprising and striking and thought a lot about it, and I also wrote about this in my notes after classroom observation. The teacher was also aware of this and a bit worried but she always tried to involve the girls and encourage them to make contributions and take part. When the teacher divided the class into two groups by gender and had just one group at a time doing different activities, both of us were surprlearnedd how all the children in both groups took an active part. It is difficult to tell, however, whether it was because of gender, because the groups were smaller (just 9 and 10 children in each group), or because of the work and the activities that took place during these lessons. In the literature there are clear examples that show that girls are more passive than boys in the classroom, with boys being much more willing to share their ideas than girls (Collins 1998; Jule, 2003). It would be interesting to follow this up with further research.

5.7 The Different Methods Used to Gain Access to Children's Ideas

The different methods used to gain access to children's ideas can also have had an effect by clarifying their ideas. When drawing, the children have to think about what they are going to draw and how they are going to represent their ideas, and when repeating the "same" drawing later the former is likely to influence the latter. The interview is also very likely to have had an effect on their ideas and influence their performance on the diagnostic tasks (learning by the interview through talking), and also the supporting material (e.g., where they had the torso in front of them and were asked about all the different organs). The results show that the children generally performed better in the diagnostic tasks than in the interview, especially on issues relating to structure, location, and the function of the stomach and the function of the lungs. In these cases, the children scored significantly higher (paired *t*-test, $p < 0.05$) on the diagnostic tasks than on the interviews. These results may suggest that the individual interview had an effect on the children's performance, which means they learned from the interview and therefore did better on the diagnostic tasks. In that case, it can be suggested that in the individual interview with the teacher, where a model of the body was used in the discussion, there was a positive effect on children's knowledge and performance.

Each of the methods used to get access to children's ideas has its strengths and weaknesses. It has already been discussed that the class discussion usually just gives ideas from a sample of children, the active children. However, although the semi active and especially the quiet children do not take part in the discussion, they seem to learn from it as they listen to the discussion and the ideas presented. The results of this research has proved that using drawings to get access to children's ideas can be a very effective way, although it has to be borne in mind that the fine motor skills of children at this age can be very differently developed, and some children have difficulties making drawings that represent their ideas. Thus, drawings alone can have considerable limitations as measurement tools. This was borne out in a few cases where the drawings did not at all represent the child's ideas and knowledge, as demonstrated by the other methods. Here, the interview proved to be a better way to get access to children's ideas. In other cases, the drawings proved to be very good to get access to and evaluate the ideas and the development of ideas. Drawings proved to be a good way for getting access to the ideas of the quiet children, which I think is very important to have in mind when assessing children's ideas and progress. However, if the drawings the children in this study made had been followed up individually by asking each child to explain their drawing, it is likely that a different and a fuller view of their ideas, as expressed in the drawings, would have been attained because the drawings sometimes were more symbolic than realistic. This is also a view expressed by Carvalho et al. (2004), but they claim the drawings young children make to represent their ideas about digestion are more symbolic, rather than realistic, and I have already mentioned a few examples that support this view in my study. Therefore, teachers have to be careful when analyzing children's drawings and use other methods along with the drawings to enable them to get a clearer picture of the children's ideas.

As the results show, the ideas the children in the study had about the body changed according to their drawings, but, as discussed before, the drawings did not always represent their knowledge and understanding. Sometimes it was as if they used the drawings as symbols to illustrate their ideas, and if there had not been other methods used to get access to their ideas we would not have a valid picture of their ideas, knowledge, and understanding about the body. But, it can be difficult to know if there is a difference between an understanding of an idea and the rote presentation, or an imitation. It should also be kept in mind that post-teaching drawings could be showing imitation (i.e., from the textbook) rather than understanding, but, it is difficult to analyze the potential difference between understanding and rote learning and the problem in teasing these apart in the data, and this holds both for the drawings and the explanatory data.

5.8   Methodological Strengths and Weaknesses

In the previous section, some of the strengths of the methods used to get information about the children's ideas about their body have been discussed. Also discussed, were some things that should be taken into consideration, such as using more than one method to get information, drawings alone giving limited information, and also the fact that information about the function of an organ and of its processes can not be seen from drawings—making it difficult to see if there is a correlation between a content of a drawing and knowledge and understanding. Therefore, it is important to use "following up" interviews where children are interviewed about their drawings. The children's drawings were used in the interview at the end of the project as a base for discussion but since there was almost a year since the children had made some of the drawings they did not give relevant information, although the interviews did in some cases clarify ideas presented in some drawings, like in the case of Óli when he drew pluses (+++) all around the "baby" in the mommy's tummy and said when asked in the interview that the pluses meant oxygen. However, it would have been better to interview the children about the drawings soon after they made them as suggested by Guichard (1995).

The different methods used to obtain information about children's ideas, that is, drawings, interviews, and diagnostic tasks proved to be very useful as is shown on Figure 4.17, panels a, b, c, and d. Considerable differences between the methods used can be seen. It has, however, to be taken into consideration that the difference may be attributed to either the setting or the operation of the scales used to analyze the drawings, or due to the sequence in which the tasks were presented, rather to the method per se. Thus, a more rigorous design would have led to more unambiguous conclusions. For instance, Figure 4.17, panels a, b, c, and d show the results for drawings made at the early stages of the study, but the interviews and the tasks were undertaken at the end of the project. It would have been relevant to ask the children to make drawings of the skeleton and the organs of the body at the end of the project too.

From the study, it is hard to conclude which teaching method is "the best one" for the individual child. The results show that for the group, demonstration and hands-on activities, along with information/telling and class discussion, were effective. I could have better probed the children in the interview about the teaching methods and more about the teaching, that is, what they thought they learned most from.

It is now clear that there are several aspects regarding children's activities and their performance that could be improved upon and certainly topics for further study.

The strength of the study is that its point of departure is the classroom, the children, and the teacher, and the interaction and how this relates to the development of the children's ideas about the body. The strength is also reflected in the breath of the study relating to children's ideas about the body and its variety in relation to structure, location, and function of bones and organs. It also shows how both the children and the teacher use the teaching materials and how this is reflected in their learning. Also shown is the considerable individual differences and how this interacts with both the teaching and the learning and how different methods show different aspects of the children's knowledge and how these interact with individual differences. It also shows how scales can be developed for summative purposes but much more importantly as an integral part of the formative dynamics.

5.9 The Contribution of This Research to the Educational Field

The research has clarified many important issues relating to the development of children's ideas about the human body, and about how these can be taught and assessed, both for formative and summative purposes.

The research has demonstrated important individual differences as referred to individual state of activity and how these may interact with the children's learning, and how both teaching and assessment may be individualized. Reykjavik Educational Authority has recently put forward a policy (Reykjavíkurborg, Menntasvið, 2006) that emphasizes "individualized learning" so this research should be a valid contribution to developing such learning. It emphasizes the importance of gauging children's initial knowledge on which to build their education. It emphasizes the importance of noting how the considerable individual differences may interact with the children's mode of learning. In Iceland, we have lacked research on learning and teaching in the primary school so this research fills a gap, but it also contributes to educational research in general as it gives valuable insight into children's ideas and how and under what conditions they develop. Furthermore, it shows some important factors that changes their ideas. It provides an important distinction between active and the passive children and their learning, and how children can be learning without showing any of the commonly assumed visual signs taken to be a prerequisite of learning, like taking part in discussion and sharing ideas. It gives information for teachers about how the different teaching methods work and what teaching methods seem to be more effective than others and how important it is to use different methods to get access to children's ideas. It also shows how important it is to use different types of assessment or measurement devices such as drawings, talking to the children, and using completion tasks. It is also of value for writers of curriculum guidelines, and writers and illustrators of teaching material, as the results show how pictures, drawings, and the text used in textbooks has an effect on children's ideas and how the curriculum materials interact in different ways with different contents.

There are, however, many issues that have arlearnedn from the study that require further study, such as the gender difference, the influence the Icelandic language has on ideas about the body compared to other languages, and how the metaphors adults use to explain issues related to the structure, function, and the processes of the different organs is likely to influence children's ideas. It would be interesting to see if children from the countryside are more familiar with the structure and the function of the organs than are children in Reykjavík. It is also important to look more closely at the influence teaching material and pictures in textbooks have on the development of ideas, and to look at the influence a teaching method like drama has on children's understanding when used on its own, because in this study drama was always used with other methods.

Not least, it would be interesting to do further research on the different methods used to get access to children's ideas and also study in more detail the passive, quiet children and their learning, and also the active children and why they are not gaining as much as the others. Thus, the research gives answers to important questions and issues but it also ralearneds new questions that give opportunities for further educational research.

5.10   Recommendation for Teaching About the Body

In light of the results of the research and my experience in doing the study, some recommendations for teaching about the body are suggested:

1. It is important to start with a whole class discussion about the human body, in order to get information about the children's existing ideas about the body, even though such information will only come from a sample of children, those happy to contribute out loud in a group setting. However, other methods should be used too to get access to ideas. Ask the children to make drawings of their ideas, the whole body, the skeleton, and the organs. If possible, the teacher should talk to the children individually, in pairs or small groups, about their drawings as the drawings may not always represent their knowledge and understanding.

2. After discussing and getting to know the pupils' ideas about the various parts of the body, it is important to pay attention to one issue at the time: bones and muscles, specific organs, how organs are connected, and how they can be part of an organ system as recommended by Reiss and Tunnicliffe (2001). It is also important to distinguish the organs' structure, location, function, and processes. *The National Curriculum Guide: Natural Science* (Menntamálaráðuneytið, 1999) put forward aims and objectives that include teaching about the body, e.g.:

   a. body parts that we see (hands, legs, feet, head, hair, teeth …)
   b. skeleton and muscles
   c. cells and reproduction
   d. senses and the brain
   e. the digestion system (including the liver and kidneys)
   f. heart, lungs, and blood circulation

But, for each part, these different aspects should be attended to when possible:

1. The teacher should use a variety of teaching methods in her teaching about the body such as hands-on activities, demonstration, drama, etc., as different teaching methods work differently for individual children as they have different modes of learning.

2. The teacher should think about how the children are going to represent their work and findings. They can draw, write, and tell about their ideas and findings.

3. The drawings and other work the children do individually should be collected, and a portfolio made for each child that will be part of a formative assessment and enable the teacher to see the development and change represented in the drawings and the other work.

4. The teacher should ensure that the environment is stimulating and encouraging. She should get books from the library and ask the

children to bring books about the body to school, and make a display of pictures and models of the organs like the heart, the brain, the skeleton. The textbook and the teaching material used in this research had perhaps greater effect than had been expected, so it is important to decide carefully what teaching material is to be used, making sure that it matches the aims and objectives in focus.

Good organization is essential for successful teaching and learning, and where the teacher is using a variety of different teaching methods and the children are working on different activities, I suggest the following points to have in mind:

1. The teacher or a small group of children carry out the activity or demonstration while the rest of the children watch and take part in the discussion. Next time another group of children performs the activity.
2. The class is divided into groups by gender where the boys and girls work separately on an issue or a task.
3. The class is divided into groups of three or four. All the groups do the same activity at the same time and the teacher controls the work and the discussion.
4. The class is divided into groups. One group does the science activity, but at the same time the other groups are working on different tasks (not necessary science tasks) that do not need as much supervision. The teacher then gets a chance to guide and take part in the discussion with the science group and therefore gets better access to their ideas.
5. The class is divided into groups. Each group works on different science tasks at the same time and all the groups do all the tasks, one after another. This is often called "the carousel"—there are several learning centers or stations in the classroom. In small groups, students rotate, or carousel clockwise, around from one station to another. The time allowed at each station can be from 20 to 40 minutes. This kind of work demands a lot of organization by the teacher, and the children have to be accustomed to the other methods mentioned before they can work independently of the teacher in this way. The teacher has to give very clear but simple instructions at the beginning, then the children work together in each group and the teacher can go between the groups and look at how they do the activity and take part in the discussion.

All these recommendations are about teaching and active learning, with emphasis on the individual pupil within a whole class situation or an educational setting.

5.11 Conclusions

In light of the results of the study and the discussion, the following conclusions are highlighted:

The study demonstrates the complexity involved in teaching about each individual content unit, that is, the bones, the heart, lungs and the blood circulation, the stomach and the digestive process, and the brain. The complexity is due both to the concepts and individual differences of the children.

There was a great individual difference between the children, but results show that the majority were gaining, that is, learning. Thus, it can be concluded that a variety of teaching methods and the visual aids used by the teacher is in itself important for the children as a group.

All of the children mastered practically all the content, although it is not clear when. Some of them mastered the content very soon, some not until the end of the teaching period and apparently not until the end of the study project, which in this sense became a part of the teaching process.

At the end of the project the children were generally more aware of the structure, location, and the function of the different organs than they were of processes and how the organs were interrelated. They were also more aware of the digestive system than other organ systems and more aware of the digestive process than the other processes.

Using drawings to get access to children's ideas is a very effective way although it has to be borne in mind that the fine motor skills of children at this age can be very differently developed, and some children have difficulties in making drawings that represent their ideas so drawings alone can be a vague research method. The imitation effect has also to be taken into account as the drawings can present imitation rather than understanding. The ideas the children in the study had about the body changed according to their drawings but the drawings did not always represent their knowledge and understanding. Sometimes they seemed to use the drawings as symbols to illustrate their ideas and if there had not been other methods used to get access to their ideas we would not have a valid picture of their ideas, knowledge, and understanding about the body. The drawings young children make to represent their ideas about digestion tend to be more symbolic rather than realistic and few examples have already been mentioned that support this view in the study. Therefore, teachers have to be careful when analyzing children's drawings and use other methods along with the drawings to enable them to get a clearer picture of the children's ideas. The teaching material used, *Let's Look at the Body* (Óskarsdóttir & Hermannsdóttir, 2001a), obviously had an effect on the children's drawings because they tended to imitate the pictures in the book when they were asked to draw their ideas.

The results show that different teaching methods have different effects on different children. Thus, a variety of methods are important in order to ensure a rich understanding and to take the ideas further. There is no single method for teaching and learning that fits all, so science education should consist of a wide variety of teaching methods, including telling and showing (demonstrating), taking into account that every individual learner is unique.

Demonstration and discussion at the same time seems to be very effective with this age group (6 and 7 years old) and seem to be even better than hands-on activities that the children are doing on their own where it can be difficult for the teacher to follow what is going on or the discussion that takes place. However, practical activities and discussion where the teacher is watching and listening to the children or leading the activity, giving information, and asking questions, also seems to be very effective. Lecturing and telling can also work well, especially if the teacher uses something visual, like models and pictures to clarify things.

The visibly active children did not seem to learn as much as the other children from the intervention according to the drawings, which suggests that they knew most of it before the intervention occurred and they did not add as much to their knowledge as the other children. Their gain was not as great as shown by the others although their initial scoring was higher. This suggests that these children, at least, needed more individualized learning opportunities as the teaching did not add to their learning.

Although the semi active and the quiet children did not take part in the discussion, they seem to learn from it as they listen to the discussion and the ideas presented. According to the results, the quiet children were not learning less than the others and the results suggest that they were learning more (in the sense of gaining more) than the visibly active children.

The girls in general were quieter and more passive than the boys according to classroom observation. Both boys and girls seemed, however, to take generally more part when divided into groups by gender than when the whole group was together.

The results show that the children performed better in the diagnostic tasks than in the interview on issues relating to structure, location, and the function of the stomach and the function of the lungs where results from the diagnostic tasks were significantly better than from the interviews. It was concluded that individual interviews had effect on the children's performance, which means they learned from the interview and at least partly therefore did better on the diagnostic tasks. So it can be suggested that individual interviews with the teacher, where a model of the body is used in discussion, has a positive effect on children's knowledge and performance although it can be time consuming. Thus the interview may have substantial value, both as a tool for formative evaluation, and also, concurrently, as an important teaching tool.

The study has demonstrated the importance of the use of scales, such as the Reiss & Tunnicliffe scale. But it was also shown that scales refined to address specific content for different age groups may be usefully developed to be used routinely both as research tools probing different specialized areas, but also as efficient diagnostic tools for the teachers.

Thus, it can be concluded that this research has clarified a number of issues related to the development of children's ideas about the body and how these may be taught and evaluated. Furthermore, it has suggested that several of these issues need more probing, ranging from asking questions about the importance of different types of content (or the different nature of the various concepts) to the variety of methods used to evaluate the learning that has taken place. This research has also put into a perspective a number of teaching practices that seem to work well for the different content, and more importantly for different students, even though some of the details of their effect still need to be clarified. Finally, this research has demonstrated the importance of the active engagement of the students as a group, but has at the same time shown how the visible activity of individuals may not correlate to learning in the manner often presumed.

Thus, in addition to the theoretical clarification discussed in the study and the substantive practical guidance for the teaching practice suggested above, this research demands further clarification of a number of theoretical issues.

APPENDIX I

Semi-Structured Interview Schedule for the Teacher

Interview Scheme for Interviews on November 20 and 27, 2003

Geturðu lýst því hvernig þú undirbýrð kennsluna (tengsl við samkennara) þekkingafræðilegan undirbúning, kennslufræðilegur undirbúningur.
Can you explain how you prepare your lesson (cooperation with fellow teachers), in terms of content and educational preparation?

Hvað er það sem ræður skipulagningunni og kennsluaðferðunum?
What factors influence the organization and the educational strategy?

Hvernig finnst þér þekking þín á viðfangsefninu nýtast þér til að kenna þetta efni?
How do you find your prior knowledge of the subject useful as you teach this lesson?

"The Brain Controls Everything", pages 189–190
Copyright © 2016 by Information Age Publishing
All rights of reproduction in any form reserved.

Hvað gerirðu til að bæta við þig þekkingu ?
How do you engage in continuing education?

Hvernig undirbúning hefur þú úr þínu eigin námi?
How has your education prepared you for this job?

Hvaða áhrif hafa samkennarar þínir?
How do your fellow teachers influence your work?

Hvernig skilgreinir þú nám?
How do you define learning?

Hvernig getur þú metið/skynjað hvort nám á sér stað, hvort börnin eru að læra.
How can you assess whether learning is taking place in your classroom?

Hvernig reynirðu að ná til nemenda og hvernig heldur þú athygli þeirra?
How do you capture your student's attention?

Hver er tilfinnig þín fyrir virkni nemenda?
Do you find your student's engaged in the classroom?

Munur á strákum á stelpum.
Difference between the genders.

Hvað finnst þér þú hafa lært á þessu verkefni?
What have you gained by partaking in this project?

Semi-Structured Interview Schedule for the Children

Children's Ideas About the Body

Interviews With Children—Scheme

Talað verður við börnin einstaklingslega. Viðtölin fara fram í herbergi við hliðina á kennarstofunni.
The children will be interviewed in private in a room beside the teacher's lounge.

Reiknað er með að hvert viðtal taki u.þ.b. 15–20 mínútur.
Each interview is expected to take approximately 15–20 minutes.

Viðtölin verða tekin upp á myndband til greiningar eftir á.
The interviews will be videotaped for analysis at a later date.

Stuðst verður við líkan af líkamanum, líffærin, og teikningar barnanna sjálfra.
A plastic model of the body will be used as props, as well as the children's own drawings.

"The Brain Controls Everything", pages 191–192
Copyright © 2016 by Information Age Publishing
All rights of reproduction in any form reserved.

Áhersla verður lögð á útlit og hlutverk líffæranna og staðsetningu þeirra.
A special emphasis will be placed upon the location of the various organs, their role, and what they look like.

Þegar hvert barn kemur í viðtalið verður búið að taka nokkur líffæri úr líkaninu. Líffærin liggja á borði fyrir framan líkamann.
At the beginning of each interview a few organs have been removed from the model and placed on the table in front of it.

Ég mun biðja barnið að taka ákveðið líffæri upp og spyr hvort það viti hvaða líffæri þetta er og hvaða hlutverki það gegni. Síðan mun ég biðja barnið að raða líffærunum á réttan stað í líkamanum og mun á meðan ræða við barnið um hlutverk líffæranna og hvernig þau tengjast.
I will ask the child to pick up a particular organ and identify it as well as answer questions about its function. I will then ask the child to put the organs in the right places on the model while I talk to the child about the role of the organs and their connections.

Síðan mun ég fara yfir teikningarnar og reyna að fá barnið að meta hvað það hefur lært. Skoða með því muninn á teikningunum.
I will then study the pictures and help the child to assess what he or she has learned by examining together the difference between the pictures.

Ég mun síðan styðjast við gátlista (sjá gátlista)
Finally, I will use checklists.

Extra Questions

Veistu hvernig við stækkum? (frumur skipta sér)
Do you know why we grow? (cells divide)

Er einhver í bekknum þínum sem veit mikið um líkamann?
Is there anyone in your class who knows a lot about the body?

Áttu einhverja bók um líkamann heima?
Do you have a book about the body in your house?

Form for Analyzing Children's Interviews, February 2004

Children's Ideas About the Body

Name _____

How babies are made/cells

Knows the structure of:

- ☐ Heart
- ☐ Lungs
- ☐ Stomach
- ☐ Intestine
- ☐ Colon
- ☐ Brain
- ☐ Ribs

Knows the position of:

- ☐ Heart
- ☐ Lungs
- ☐ Stomach
- ☐ Intestine
- ☐ Colon
- ☐ Brain
- ☐ Ribs

"The Brain Controls Everything", pages 193–194
Copyright © 2016 by Information Age Publishing
All rights of reproduction in any form reserved.

Knows the function of:

☐ Heart

☐ Lungs

☐ Stomach

☐ Brain

Knows the way the food goes from mouth to anus

Knows liver/kidneys

Other relevant knowledge:

APPENDIX **IV**

Semi-Structured Interview Schedule for the Parents, March 2004

Guidelines for Parent Interviews

Hefur barnið haft sérstakan á huga á líkamanum, líffærunum, beinunum?
Has the child shown special interest in the body, particular organs, or bones?

Hefur barnið spurt mikið um líkamann, hvernig hann er og til hvers líffærin eru?
Has the child asked questions about the body, how it functions, and the role of the various organs?

Hefur hlutverk einhverra sérstakra líffæra verið útskýrt fyrir barninu?
Has the role of particular organs been explained to the child?

Er til einhver bók á heimilinu um líkamann sem barnið hefur skoðað mikið?
Is there a book on the body in the home that has caught the interest of the child?

"The Brain Controls Everything", pages 195–196
Copyright © 2016 by Information Age Publishing
All rights of reproduction in any form reserved.

Hefur barnið horft á eitthvert sérstakt sjónvarpsefni eða videóspólu þar sem fjallað er um líkamann.
Has the child watched a special television program, a DVD, or a video tape about the body?

Veistu hvort fjallað um líkamann sérstaklega á leikskóla barnsins?
Are you aware of any specified instruction on the body at the child's nursery school?

Hefur vinnan um líkamann í skólanum skilað sér heim?
Has the child brought up the subject of the workshop on the body that the child took part in at school?

Hefur barnið beinbrotnað eða verið á sjúkrahúsi?
Has the child suffered broken bones or ever been hospitalized?

Form for Information Collected From Children's Drawings

Children's Ideas About the Body

Name _____

Knows the structure of:

☐ Heart
☐ Lungs
☐ Stomach
☐ Intestine
☐ Colon
☐ Brain

Knows the position of:

☐ Heart
☐ Lungs
☐ Stomach
☐ Intestine
☐ Colon
☐ Brain

"The Brain Controls Everything", pages 197–198
Copyright © 2016 by Information Age Publishing
All rights of reproduction in any form reserved.

Knows the function of:

☐ Heart

☐ Lungs

☐ Brain

Knows the way the food goes from mouth to anus

Knows liver/kidneys

Other relevant knowledge:

Diagnostic Tasks

I. One task involved drawing a line between a name of an organ and the function of it.
II. Another task included statements, and the children were asked to write T (True) and F (False) accordingly.
III. The third task involved coloring organs in certain colors.
IV. In the fourth task the children were asked to draw with a red color the way the food travels from the mouth and through the body.
V. In the fifth task they were asked to connect the words to the appropriate spots on a drawing of the skeleton.
VI. The last two (VIa and b) were questions and answers to tasks.

"The Brain Controls Everything", pages 199–206
Copyright © 2016 by Information Age Publishing
All rights of reproduction in any form reserved.

Tasks I and II

The Body

Name: _____

Date: _____

Connect the Terms With the Right Definition

Hjartað The heart	*stjórnar öllum líkamanum* controls the body
Lungun The lungs	*flytja boð til heilans* transmit messages to the brain
Taugafrumur The nerve cells	*sjá okkur fyrir súrefni* provide the oxygen supply
Heilinn The brain	*dælir blóði* pumps blood
Rifbeinin/the ribs The ribs	*verja hjartað og lungun* protect the heart and the lungs

True or False (Write T or F as Appropriate)

	True	False
Lungun stjórna öllum líkamanum. The body is controlled by the lungs.	☐	☐
Vöðvarnir hjálpa okkur að hreyfa líkamann. The muscle enable the body to move.	☐	☐
Blóðið flyst með æðunum um allan líkamann. The blood is carried throughout the body through the vains.	☐	☐
Taugafrumurnar bera boð til heilans. The nerve cells transmit messages to the brain.	☐	☐
Rifbeinin verja heilann. The ribs protect the brain.	☐	☐
Þegar sæðisfruma frá pabbanum sameinast *eggfrumu frá mömmunni verður til barn.* When the father's sperm joins the mother's egg a baby is created.	☐	☐

Task III

Organs

Name: _____

Date: _____

Skoðaðu myndina vel.
Take a good look at the picture.

Litaðu **hjartað** *rautt.*
Color the heart red.

Litaðu **lungun** *blá.*
Color the lungs blue.

Litaðu **lifrina** *brúna.*
Color the liver brown.

Litaðu **magann** *gulan.*
Color the stomach yellow.

Litaðu **ristilinn** *appelsínugulan.*
Color the colon orange.

Litaðu **Þarmana** *græna.*
Color the intestine green.

Litaðu **leilann** *svartan.*
Color the brain black.

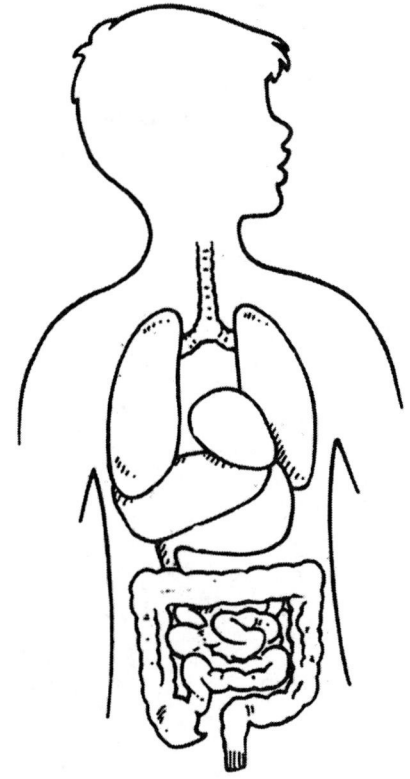

Task IV

Digestion

Name: _____

Date: _____

Teiknaðu með rauðum lit leiðina sem maturinn fer frá munninum og út úr líkamanum.

Draw in red the passage of the food from the mouth through the body.

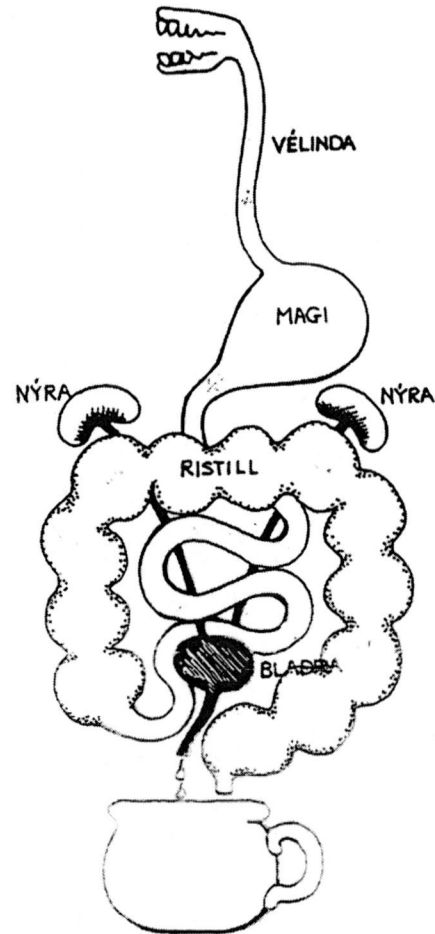

Graphic by Ragnheiður Gestsdóttir.

Task V

The Skeleton

Name: _____

Date: _____

Tengdu orðin við rétta staði á myndinni.
Connect the words to the appropriate spots on the picture.

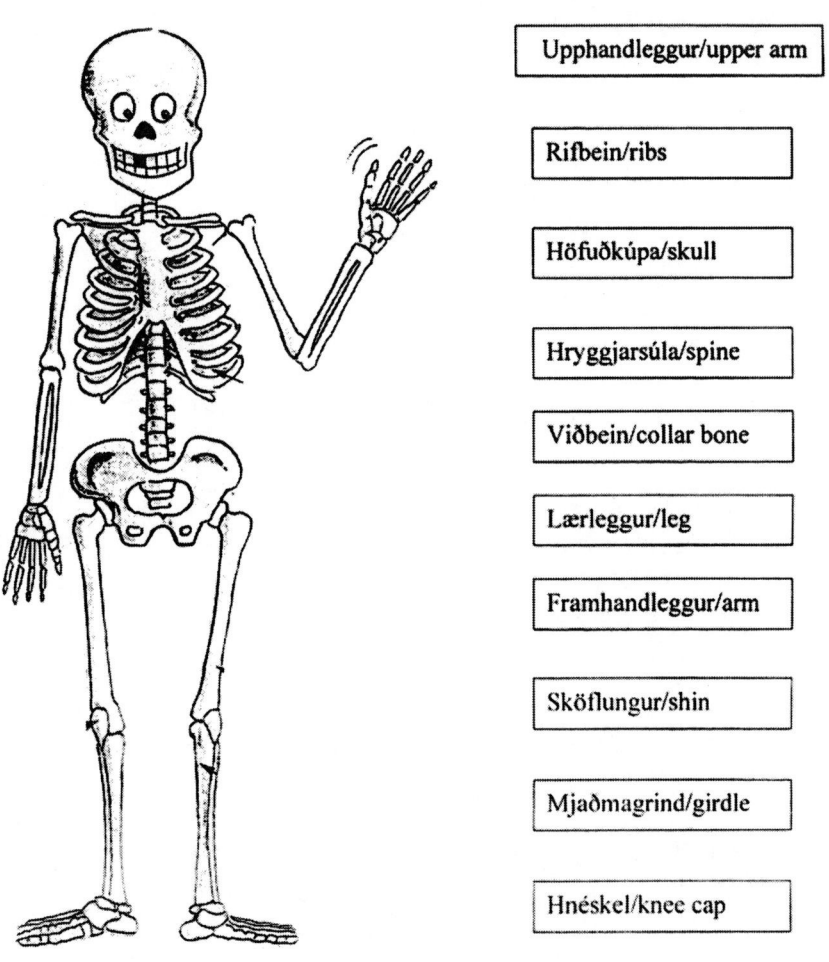

Upphandleggur/upper arm

Rifbein/ribs

Höfuðkúpa/skull

Hryggjarsúla/spine

Viðbein/collar bone

Lærleggur/leg

Framhandleggur/arm

Sköflungur/shin

Mjaðmagrind/girdle

Hnéskel/knee cap

Task VIa and b

The Body

Task VIa—Riddles

Name: _____

Date: _____

Ég slæ hraðar þegar þú hleypur.
Ég dæli blóði um allan líkamann.
Ég er vöðvi.
Hver er ég? _____

I beat faster when you run.
I pump blood throughout the body.
I am a muscle.
Who am I?_____

* * *

Ég tek við upplýsingum frá augum, eyrum, húð og vöðvum.
Ég get varað þig við hættum.
Höfuðkúpan verndar mig
Hver er ég? _____

Protected by the skull, I receive
data from eyes, ears, skin and muscles.
I can warn against danger.
Who am I ? _____

* * *

Á ferðalaginu um líkamann fer blóðið til okkar og nær í súrefni.
Við erum tvö.
Þú andar með okkur.
*Hver erum við?*_____

On its tour around the body the blood makes a stop
and receives oxygen from the two of us.
You breathe with us.
Who are we? _____

* * *

*Ég er eins og poki úr teygjanlegu efni sem vinnur líkt
og hrærivél eða matvinnsluvél.
Inni í mér blandast fæðan þangað til hún er orðin
að graut eða mauki.
Hver er ég?* _____

I am like a bag made of elastic material operating
in ways similar to a mix master or a food processor.
The food is mixed up inside of me until it has the
consistency of a pudding or puree.
Who am I ? _____

* * *

*Við sendum boð til og frá heilanum.
Við erum eins og símalínur, ring, ring ...
Hverjar erum við?* _____

We send data to the brain and back..
We are like phone lines, ring, ring...
Who are we ?_____

Task VIb—Questions

Name: _____

Date: _____

Answer the Following Questions:

1. Hvaða mikilvæga líffæri verndar höfuðkúpan?
   Which vital organ is protected by the skull?

2. Hvað líffæri dælir blóðinu um líkamann?
   Which organ pumps the blood through the body?

3. Hvað gera rauðu blóðkornin?
   What role do the red blood cells perform?

4. Hvað gera hvítu blóðkornin?
   What role do the white blood cells perform?

5. Hvaða frumur eru það sem flytja skilaboð til heilans?
   Which cells transmit messages to the brain?

6. Hvað heita beinin sem verja hjartað og lungun?
   What are the bones called which protect the heart and the lungs?

7. Á milli flestra beina eru _____ sem gera okkur
   kleift að hreyfa okkur.
   *Between most bones there are _____ which enable
   the movement of the body.*

Form For Information Collected From Children's Diagnostic Tasks, February 2004

Children's Ideas About the Body

Name _____

Reproduction

Knows the structure of:

- ☐ Heart
- ☐ Lungs
- ☐ Stomach
- ☐ Intestine
- ☐ Colon
- ☐ Brain
- ☐ Liver

Knows the position of:

- ☐ Heart
- ☐ Lungs
- ☐ Stomach
- ☐ Intestine
- ☐ Colon
- ☐ Brain
- ☐ Liver

"The Brain Controls Everything", pages 207–208
Copyright © 2016 by Information Age Publishing
All rights of reproduction in any form reserved.

Knows the function of:

- ☐ Heart

- ☐ Lungs

- ☐ Muscles

- ☐ Brain

- ☐ Ribs/Scull

Knows the way the food goes from mouth to anus

The function of the stomach

Knows liver/kidneys

Other relevant knowledge:

References

Alexander, R. (2004). *Towards dialogic teaching: Rethinking classroom talk.* Cambridge, England: Dialogos.

Appleton, K., & Asoko, H. (1996). A case study of a teacher's progress toward using a constructivist view of learning to inform teaching in elementary science. *Science Education, 80*(2), 165–180.

Banister, P., & Parker, I. (1994). *Qualitative methods in psychology: A research guide.* Philadelphia, PA: Open University Press.

Barnes, D. (1976). *From communication to curriculum.* Middlesex, England: Penguin Books.

Bernstein, A. C., & Cowan, P. A. (1975). Children's concepts of how people get babies. *Child Development, 46,* 77–91.

Black, P. & Harlen, W. (1995). *Living processes. Teachers guide.* London, England: London Educational.

Bleach, K. (Ed.). (1998). *Raising boys' achievement in schools.* Stoke on Trent, England: Trentham Books.

Bogdan, C. R., & Biklen, S.K. (2003). *Qualitative research for education.* Boston, MA: Allyn and Bacon.

Boshell, M. (1995). *What I learned from giving quiet children space.* Aston University: Language Studies Unit.

Bruner, J. S. (1996). *The culture of education.* Cambridge, MA: Harvard University Press.

Bruner, J. S., & Haste, H. (1987). *Making sense: The child's construction of the world.* London, England: Methuen.

Burman, E., & Parker, I. (1993). *Discourse analytic research.* London, England: Routledge.

"The Brain Controls Everything", pages 209–217
Copyright © 2016 by Information Age Publishing
All rights of reproduction in any form reserved.

Burns, C., & Myhill, D. (2004). Interactive or inactive? A consideration of the nature of interaction in whole class teaching. *Cambridge Journal of Education, 34*(1), 35–49.

Cangelosi, J. S. (1992). *Systematic teaching strategies.* New York, NY: Longman.

Carey, S. (1985). *Conceptual change in childhood.* Cambridge, MA: MIT Press.

Caroll, R.T. (Ed.). (2005). *Skeptic's Dictionary.* Retrieved May 24, 2006, from http://skepdic.com/vitalism.html

Carugati, F. (1999). From Piaget and Vygotsky to learning activities: A long journey and an inescapable issue. In M. Hedegaard & J. Lompscher (Eds.), *Learning activity and development* (pp. 211–234). Aarhus, Denmark: Aarhus University Press.

Carvalho, G. S., Silva, R., Lima, N., & Coquet, E. (2004). Portuguese primary school children´s conceptions about digestion: Identification of learning obstacles. *International Journal of Science Education, 26*(9), 1111–1130.

Chi, M., Siler, S., Jeong, H., Yamauchi, T., & Haussmann, R. (2001). Learning from human tutoring. *Cognitive Science, 25,* 471–533.

Clément, P. (2003). *Situated conception and obstacles: The example of digestion and excretion.* Dordrecht: Kluwer Academic.

Cohen, L., Manion, L., & Morrison, K. (2000). *Research methods in education* (5th ed.). London, England: RoutledgeFalmer.

Cole, J. (1996) *The Magic School Bus. Inside the human body.* London, England: Scholastic.

Cole, M., & Wertsch, J. V. (1996). Beyond the individual-social antimony in discussions of Piaget and Vygotsky. *Human Development, 39*(5), 250–257.

Collins, J. (1996). *The quiet child.* London, England: Cassell.

Collins, J. (1997, September). *Barriers to Communication in Schools.* Paper presented at the British Educational Research Association Annual Conference, York, England.

Collins, J. (1998, August). *Playing truant in mind: The social exclusion of quiet pupils.* Paper presented at the British Educational Research Association Annual Conference, Belfast, Ireland.

Corbin, J., & Strauss, A. (1996) *Basics of Qualitative Research.* London, England: Sage.

Cornu, R. L., Peters, J., & Collins, J. (2003, September). *What are the characteristics of constructivist learning cultures?* Paper presented at the British Educational Research Association Annual Conference, Edinburgh, Scotland.

Cox, M., Perara, J., Koyasu, M., & Hiranuma, H. (2001). Children's human figure drawings in the UK and Japan: The effects of age, sex and culture. *The British Psychology Society, 19,* 275–292.

Creswell, J. W. (1998). *Qualitative inquiry and research design: Chosing among five traditions.* Thousand Oaks, CA: Sage.

Cuthbert, A. (2000). Do children have a holistic view of their internal body maps? *School Science Review, 82,* 25–32.

Dalgarno, B. (2001). Interpretations of constructivism and consequences for computer assisted learning. *British Journal of Educational Technology, 32*(2), 183–194.

Daniels, H. (2001). *Vygotsky and pedagogy*. London, England: RouthledgeFalmer.

Delamont, S. (2002). *Fieldwork in educational settings*. London, England: Routledge.

Denzin, N. K., & Lincon, Y. S. (1998). *Strategies of qualitative inquiry*. Thousand Oaks, CA: Sage.

Driver, R. (1983). *The pupil as scientist?* Milton Keynes: Open University Press.

Driver, R., Guesne, E., & Tiberghien, A. (1985). *Children's ideas and the learning of science*. Milton Keynes, England: Open University Press.

Driver, R., Squires, A., Rushworth, P., & Wood-Robinson, V. (1994). *Making sense of secondary science*. London, Englan: RoutledgeFalmer.

Elstgeest, J. (1985a). Encounter, interaction, dialogue. In W. Harlen (Ed.), *Primary science. Taking the plunge* (pp. 9–20). Oxford, England: Heinemann Educational.

Elstgeest. J. (1985b). The right question at the right time. In W. Harlen (Ed.), *Primary science. Taking the plunge* (pp. 36–46). Oxford, England: Heinemann Educational.

Emerson, R. M., Fretz, R. I., & Shaw, L. L. (1995). *Writing ethnographic fieldnotes*. Chicaco, IL: The University of Chicaco Press.

Farmery, C. (2002). *Teaching science 3–11*. London, England: Continuum.

Farmery, C. (2005). *Getting the buggers into science*. London, England: Continuum.

Fisher, R. (2005). *Teaching children to learn* (2nd ed.). Cheltenham, England: Nelson Thornes.

Fontana, A. (1994). *Ethnographic trends in the postmodern era*. London, England: The Guildford Press.

Fosnot, C. T. (1996). *Constructivism: Theory, perspectives and practice*. New York, NY: Teachers College Press.

Foster, P. (1996). *Observing schools*. London, England: Paul Chapman.

Frost, J. (1997). *Creativity in primary science*. Buckingham, England: Open University Press.

Gellert, E. (1962). Children's conceptions of the content and functions of the human body. *Genetic Psycology Monographs, 65*, 293–405.

Gillham, B. (2000). *The research interview*. London, England: Continuum.

Goldman, R., & Goldman, D. (1982). How children perceive the origins of babies and the roles of mothers and fathers in procreation: A cross-national study. *Child Development, 53*, 491–504.

Gould, J. S. (1996). *A constructivist perspective on teaching and learning in the language art*. New York, NY: Teachers College Press.

Graesser, A. C., Person, N., & Magliano, J. (1995). Collaborative dialog patterns in naturalistic one-to-one tutoring. *Applied Cognitive Psychology, 9*, 359–387.

Graue, M. E., & Walsh, D. J. (1998). *Studying children in context*. Thousand Oaks, CA: Sage.

Greig, A., & Taylor, J. (1999). *Doing research with children*. London, England: Sage.

Guichard, J. (1995). Designing tools to develop the conception of learners. *International Journal of Science Education, 17*(2), 243–253.

Gurian, M. (2001). *Boys and girls learn differently! A guide for teachers and parents*. San Francisco, CA: Jossey-Bass.

Gustavson, A. (1996). *Three basic steps in formal data structure analysis*. Unpublished manuscript, Stockholm, Sweden: University of Stockholm, Department of Education.

Haney, W., Russel, M., & Bebell, D. (2004). Drawing on education: Using drawings to document schooling and support change. *Harward Educational Review, 74*(3), 241–271.

Harlen, W. (1992). *The teaching of science*. London, England: David Fulton.

Harlen, W. (1993). *Teaching and learning primary science* (2nd ed.). London, England: Paul Chapman.

Harlen, W. (2000). *The teaching of science in primary schools* (3rd ed.). London, England: David Fulton.

Harlen, W. (2004a). Questions and dialogue. *Primary Science Review, 83*, 2–3.

Harlen, W. (2004b). Talking and writing, have we got the balance right? *Primary Science Review, 83*, 17–19.

Harlen, W. (2006). *Teaching, learning and assessing science 5–12* (4th ed.). London, England: Sage.

Harlen, W., Macro, C., Reed, K., & Schilling, M. (2003). *Making progress in primary science*. London, England: RouthledgeFalmer.

Harlen, W., & Qualter, A. (2004). *The teaching of science in primary school* (4th ed.). London, England: David Fulton.

Hatano, G., & Inagaki, K. (1994). Young children's naive theory of biology. *Cognition, 50*, 171–188.

Hatano, G., & Inagaki, K. (1997). Qualitative change in intuitive biology. *European Journal of Psychology of Education, XII*(2), 111–130.

Hatch, J. A. (2002). *Doing qualitative research in education settings*. New York, NY: State University of New York Press.

Hayes, D. (2004). *Foundations of primary teaching* (3rd ed.). London, England: David Fulton.

Hedegaard, M. (1999). The influence of societal knowledge traditions on children's knowledge traditions on children's thinking and conceptual development. In M. Hedegaard & J. Lompscher (Eds.), *Learning activity and development* (pp. 22–50). Aarhus, Denmark: Aarhus University Press.

Helldén, G. (1999). A longitudinal study of students' understanding of conditions for life, growth and decomposition. In M. Bandiera, S. Caravita & S. Torracca (Eds.), *Research in science education in Europe* (pp. 23–30). Dordrecht, The Netherlands: Kluwer Academic.

Helldén, G. (2004a). 19-åriga elevers uppfattningar om utveclingen av sitt eget lärande om bioloiska fenomen. In E. Hendriksen & M. Ödegaard (Eds.), *Naturfagens didaktikk–en disiplin i forandring? Det 7. Nordiske forskersymposiet om undervisning i naturfag i skolen* (pp. 315–327). Kristansand, Norway: Höyskole Forlaget.

Helldén, G. (2004b). En longitudinell studie av hut lärande i naturvetenskap utvecklas tidigt i grundskolan. In E. Hendriksen & M. Ödegaard (Eds.), *Naturfagens didaktikk–en disiplin i forandring? Det 7. Nordiske forskersymposiet om undervisning i naturfag i skolen* (pp. 301–314). Kristiansand, Norway: Höyskole Forlaget.

Heyes, D., Symington, D., & Martin, M. (1994). Drawing during science activity in the primary school. *International Journal of Science Education, 19,* 93–105.

Hill, M. (1997). Participatory research with children. *Child and Family Social Work, 2,* 171–183.

Hodson, D., & Hodson, J. (1998). From constructivism to social constructivism: A Vygotskian perspective on teaching and learning science. *School Science Review, 79*(289), 33–41.

Holbrook, H. T. (1987). The quiet student in your classroom. *Language Arts, 64*(5) 554–557.

Holgersson, I. (2003). *Children's use of metaphors and analogies while explaining matter of transformations.* Retrived March 15, 2005, from http://www1. phys.uu.nl/esera2003/programme/pdf%5C162S.pdf

Holgersson, I. (2004). Barns uppfattning av materia–varför blir ett ljus mindre när det brinner? In E. Hendriksen & M. Ödegaard (Eds.), *Naturfagenes didaktikk–en disiplin i forandring? Det 7. Nordiske forskersymposiet om undervisning i naturfag i skolen* (pp. 329–344). Kristiansand, Norway: Höyskole Forlaget.

Hollins, M., Whitby, V., Lander, L., Parson, B., & Williams, M. (1998). *Progression in primary science.* London, England: David Fulton.

Hu, Y., & Fell-Eisenkraft, S. (2003). Immigrant Chinese students' use of silence in the language arts classroom: Perceptions, reflections, and actions. *Teaching and Learning, 17*(2), 55–65.

Hyun, E. (2005). A study of 5- to 6-year old children's peer dynamics and dialectical learning in a computer based technology-rich classroom environment. *Computers and Education, 44,* 69–91.

Inagaki, K., & Hatano, G. (1993). Young children's understanding of the mind-body distinction. *Child Development, 64,* 1534–1549.

Jaakkola, R., O., & Slaughter, V. (2002). Children's body knowledge: Understanding 'life' as a biological goal. *British Journal of Developmental Psycology, 20,* 325–342.

Janesick, V. (2003). *The choreography of qualitative research design* (2nd ed.). Thousand Oaks, CA: Sage.

Jule, A. (2003, September). *Girls' talk; Girls' silence.* Paper presented at the British Educational Research Association Annual Conference, Edinburgh, Scotland.

Kaufman, J. (1976). *Svona erum við.* Reykjavík, Iceland: Setberg.

Keogh, B., & Naylor, S. (1996, September). *Teaching and learning in science: A new perspective.* Paper presented at the British Educational Research Association Annual Conference, Lancaster, England.

Keogh, B. & Naylor, S. (2004). Children's ideas, children's feelings. *Primary Science Review, 82*, 18–20.

Kozulin, A. (2003). Psychological tools and mediated learning. In A. Kozulin, B. Gindis, V. Ageyev & S. Miller (Eds.), *Vygotsky's educational theory in cultural context* (pp. 15–38). Cambridge, MA: Cambridge University Press.

Kutnick, P. (2001). Expanding the predominant interpretation of ZPD in schooling contexts: Learning and mutuality. In M. Hedegaard (Ed.), *Learning in classroom* (pp. 77–120). Aarhus, Denmark: Aarhus University Press.

Kvale, S. (1996). *Interviews*. London, England: Sage.

Kyriacou, K. (1997). *Effective teaching in schools, theory and practice* (2nd ed.). Cheltenham, England: Nelson Thornes.

Lawson, A. E. (1988). The acquisition of biological knowledge during childhood: Cognitive conflict or tabula rasa? *Journal of Research in Science Teaching, 25*, 185–199.

Lofland, J., & Lofland, L. H. (1995). *Analyzing social settings. A guide to qualitative observation and analysis* (3rd ed.). Belmont, CA: Wadsworth.

Lomangino, A. G., Nicholson, J. & Sulzby, E. (1999). The influence of power relations and social goals on children's collaborative interactions while composing on computer. *Early Childhood Research Quarterly, 14*(2), 197–228.

Marinosson, G. (1998). The ethnographic approach. *Educational and Child Psychology, 15*(3), 34–43.

Mauthner, M. (1997). Methodological aspects of collecting data from children: Lessons from three research projects. *Children & Society, 11*, 16–28.

Maynard, T. (2001). The student teacher and the school community of practice: A consideration of "learning as participation". *Cambridge Journal of Education, 31*(1), 44–49.

McCroskey, J., C. (1980). Quiet children in the classroom: On helping not hurting. *Communication Education, 29*, 239–244.

Menntamálaráðuneytið. (1999). *Aðalnámskrá grunnskóla, náttúrufræði (National Curriculum Guide: Natural Science)*. Reykjavík, Iceland: Menntamálaráðuneytið (The Ministry of Education).

Mercer, N. (1995). *The guided construction of knowledge*. Clevedon, England: Multilingual Matters.

Mercer, N., Wegerif, R., & Dawes, L. (1999). Children's talk and the development of reasoning in the classroom. *British Educational Research Journal, 25*(1), 95–111.

Merriam, S. B. (1988). *Case study research in education: A qualitative approach*. San Fransisco, CA: Jossey-Bass.

Miles, M., & Hubermann, M. (1994). *Qualitative data analysis: An expanded sourcebook* (2nd ed). Thousand Oaks, CA: Sage.

Miller, J., & Bartsch, K. (1997). The development of biological explanation: Are children vitalists? *Developmental Psychology, 33*(1), 56–64.

Mortimer, E. F., & Scott, P. H. (2003). *Meaning making in secondary science classrooms*. Maidenhead, England: Open University Press.

Myhill. D. (2003). Principled understanding? Teaching the active and passive voice. *Language and Education, 17*(5), 355–369.

Myhill, D., & Brackley, M. (2004). Making connections: Teachers' use of children's prior knowledge in whole class discourse. *British Journal of Educational Studies, 52*(3), 263–275.

Naylor, S., & Keogh, B. (2000). *Concept cartoons in science education.* Sandbach, England: Millgate House.

Noddings, N. (1995). *Philosophy of education.* Colorado, CO: Westview Press.

Norðdahl, K. (2004). Förskolebarns ideer om naturen. In E. Hendriksen & M. Ödegaard (Eds.), *Naturfagenes didaktikk–en disiplin i forandring? Det 7. Nordiske forskersymposiet om undervisning i naturfag i skolen* (pp. 387–397). Kristiansand, Norway: Höyskole Forlaget.

Ogborn, J., Kress, G., Martins, I., & McGillicuddy. (1996). *Explaining science in the classroom.* Maidenhead, England: Open University Press.

Ogden, L. (2000). Collaborative tasks, collaborative children: An analysis of reciprocity during peer interaction at key stage 1. *British Educational Research Journal, 26*(2), 211–226.

Ormrod, J. E. (1995). *Educational psychology, principles and applications.* Upper Saddle River, New Jersey: Prentice Hall.

Osborne, J., Wadsworth, P., & Black. (1992). *Processes of life: Primary SPACE project research report.* Liverpool, England: Liverpool University Press.

Osborne, J. (1996). Beyond constructivism. *Science Education, 80*(1), 52–82.

Osborne, M. D. (1997). Balancing individual and the group: A dilemma for the constructivist teacher. *Journal of Curriculum Studies, 29*(2), 183–196.

Osborne, R., & Freyberg, P. (1985). *Learning in science.* Auckland: Heinemann.

Óskarsdóttir, G., & Hermannsdóttir, R.. (2001a). *Komdu og skoðaðu likamann (Lets look at the body).* Reykjavík, Iceland: Námsgagnastofnun.

Óskarsdóttir, G., & Hermannsdóttir, R. (2001b). *Komdu og skoðaðu likamann (Lets look at the body).* Retrieved February 20, 2003 from http://www.nams.is/komdu/likamann/likaminn_frames.htm

Piaget, J. (1977). *Equilibration and cognitive structures.* New York, NY: Vikings.

Porter, J.A., & Harwood, P.J.. (2003, September). *Developing primary teachers' confidence in using constructivist approaches in science and its impact on children's understanding and attainment.* Paper presented at the British Educational Research Association Annual Conference, Edinburgh, Scotland.

Reiss, M.J. (2000). *Understanding science lessons.* Buckingham: University Press.

Reiss, M.J., & Tunnicliffe, S.D. (1999a). Children's knowledge of the human skeleton. *Primary Science Review, 60,* 7–10.

Reiss, M.J., & Tunnicliffe, S.D. (1999b). Conceptual development. *Journal of Biological Education, 34*(1), 13–16.

Reiss, M.J., & Tunnicliffe, S.D. (2001). Student's understandings of human organs and organ systems. *Research in Science Education, 31,* 383–399.

Reiss, M.J., Tunnicliffe, S.D., Andersen, A.M., Bartoszeck, A., Carvalho, G.S., Chen, SY. et al. (2002). An international study of young peoples' drawings of what is inside themselves. *Journal of Biological Education, 36*(2), 58–64.

Reykjavíkurborg, Menntasvið (2006). *Stefna Reykjavíkurborgar í menntamálum–10 ára framtíðarsýn* (Reykjavík policy in education–a 10 year plan). Retrieved September 3, 2006 from http://www.reykjavik.is/Portaldata/1/Resources/skjol/svid/menntasvid/pdf_skjol/utgafur/grunnskolar/stefnurog-stefnumorkun/menntastefa-10arasyn-rett.pdf

Richardson, I. (2006). What is good science education? In W. Harlen (Ed.), *ASE guide to primary science education* (pp. 16–24). Hatfield, England: Association for Science Education.

Richardson, L. (1994). *Writing: A method of inquiry.* Thousand Oaks, CA: Sage.

Robson, C. (2002). *Real world research.* Oxford, England: Blackwell.

Rowlands, M. (2001). The development of children's biological understanding. *Journal of Biological Education, 35*(2), 66–68.

Selley, N. (1999). *The art of constructivist teaching in the primary school.* London, England: David Fulton.

Shahrimin, M. I., & Butterworth, D. M. (2002). Young children's collaborative interactions in a multimedia computer environment. *The Internet and Higher Education, 4*, 203–215.

Shakespeare, D. (2004). Science beyond words? *Education in Science, 209*, 14–15.

Siegler, R. S. (1996). *Emerging minds, the process of change in children's thinking.* Oxford: Oxford University Press.

Slaughter, V., & Lyons, M. (2003). Learning about life and death in early childhood. *Cognitive Psychology, 46*(1), 1–30.

Solomon, J. (1994). The rise and fall of constructivism. *Studies in Science Education, 23*, 1–19.

Solomon, J., & Lee, J. (1991). *School home investigations in primary science* (Vol. 1). Hatfield: Association for Science Education.

Stake, R. E. (1998). Case studies. In N.K. Denzin & Y.S. Lincon (Eds.), *Strategies of qualitative inquiry* (pp. 86–109). Thousand Oaks, CA: Sage.

Teixeira, F. M. (2000). What happens to the food we eat? Children's conceptions of the structure and function of the digestive system. *International Journal of Science Education, 22*(5), 507–520.

Topping, K. (1992). Cooperative learning and peer tutoring: An overview. *The Psychologist, 5*, 151–161.

Toyama, N. (2000). "What are food and air like inside our bodies?" Children's thinking about digestion and respiration. *International Journal of Behavioural Development, 24*(2), 222–230.

Tunnicliffe, S. D. (2004). Where does the drink go? *Primary Science Review, 85*, 8–10.

Tunnicliffe, S. D., & Reiss, M.J. (1998). The place of living organisms in children's lives. *Short Communications,* 108–114.

Tunnicliffe, S. D. & Reiss, M.J. (1999a). Learning about skeletons and other organ systems of vertebrate animals. *Science Education International, 10*(1), 29–33.

Tunnicliffe, S. D., & Reiss, M.J. (1999b). Students' understandings about animal skeletons. *International Journal of Science Education, 21*(11), 1187–1200.

Tunnicliffe, S. D., & Reiss, M.J. (1999c). Building a model in the environment: How do children see animals? *Journal of Biological Education, 33*(3), 142–148.

Van Maanen, J. (1996). *Ethnography.* London, England: Routledge.

von Glasersfeld, E. (1996). Introduction: Aspects of Constructivism. In C.T. Fosnot (Ed.), *Constructivism: Theory, Perspectives, and Practice* (pp. 3–7). New York, NY: Teachers College Press.

Vygotsky, L. (1978). *Mind in society: The development of higher psychological processes.* Cambridge, MA: Harvard University Press.

Vygotsky, L. (1986). *Thought and language.* Cambridge, MA: The MIT Press.

Vygotsky, L. (1987). *Thinking and speech* (Vol. 1). New York, NY: Plenum Press.

Wadsworth, B. J. (1996). *Piaget's theory of cognitive and affective development: Foundations of constructivism* (5th ed.). New York, NY: Longman.

Ward, H., Roden, R., Hewlett, C., & Foreman, J. (2005). *Teaching science in the primary classroom. A practical guide.* London, England: Paul Chapman.

White, R., & Gunstone, R. (1992). *Probing understanding.* London, England: The Falmer Press.

Williams, T., Wetton, N., & Moon, A. (1989). *A way in: Five key areas of health education.* London, England: Health Education Authority.

Wilson, B. G., & Myers, K. M. (2000). Situated cognition in theoretical and practical context. In D. H. Jonassen & S. M. Land (Eds.), *Theoretical foundations of learning environments* (pp. 57–88). Mahwah, New Jersey: Lawrence Erlbaum.

Wolcott, H. F. (1994). *Transforming qualitative data: Description, analysis and interpretation.* Thousand Oaks, CA: Sage.

Wood, D., Bruner, J. S., & Ross, G. (1976). The role of tutoring in problem solving. *Journal of Child Psychology and Psychiatry, 17,* 89–100.

Yin, R. K. (2003). *Case study research: Design and methods* (3rd ed.). Thousand Oaks, CA: Sage.

CPSIA information can be obtained
at www.ICGtesting.com
Printed in the USA
FFOW01n2238110416
23139FF

9 781681 233789